SpringerBriefs in Earth System Sciences

For further volumes:
http://www.springer.com/series/10032

Michelle Goman

Human Environment Interactions – Volume 2

Reconstructing the Natural and Anthropogenic Landscape

Springer

Michelle Goman
Department of Geography and Global Studies
Sonoma State University
Rohnert Park, CA
USA

ISSN 2191-589X ISSN 2191-5903 (electronic)
ISBN 978-3-642-36879-0 ISBN 978-3-642-36880-6 (eBook)
DOI 10.1007/978-3-642-36880-6
Springer Heidelberg New York Dordrecht London

Library of Congress Control Number: 2013953590

Printed on acid-free paper

Springer is part of Springer Science+Business Media (www.springer.com)

Acknowledgments

I would like to thank the following people for their thoughtful comments on the chapters in this volume: Jessica Hedgepeth Balkin (University of Colorado, Boulder), Raymond Mueller (The Richard Stockton College of New Jersey), and Susan Zimmerman (Lawrence Livermore National Laboratories).

Contents

Contributors

Paul E. Buck is an Anthropologist and an Educator. He received his Ph.D. in Archeology from the University of Washington, and is currently a Faculty of the Desert Research Institute and Nevada State College (Nevada System of Higher Education). He has been involved in archeological and anthropological projects in a wide variety of contexts in western North America and Egypt for almost 30 years. His research interests include prehistoric human adaptation to arid environments of western North America, the transition from food collecting to food producing economies in the Southwestern U.S. and Egypt, the impact of technological change on prehistoric cultures, and applications of remote sensing and geoarcheology to prehistory. In addition to his research efforts, Buck is involved in a number of science education projects, including teaching K-12 teachers about climate change research and assisting high school students prepare competitive research projects for regional, national, and international science fairs. He is also committed to involving a greater diversity of students in math and science. As the Director of Nevada MESA Program (Mathematics Engineering Science Achievement) for 6 years, he developed programing for hundreds of under-represented students to encourage them to pursue college degrees and careers in STEM fields.

Dorothy E. Freidel is a Geoarcheologist, Geomorphologist, and Paleoenvironmentalist with a Ph.D. from the University of Oregon. Over the past 30 years, her research has focused on investigating past environments and their effects on ancient people in Oregon, Guatemala, and Ecuador during the latest Pleistocene and Holocene. She is Prof. Emerita of the Department of Geography and Global Studies at Sonoma State University, past President of the Association of Pacific Coast Geographers, and served as Editor of The California Geographer from 2004 to 2010. One of her most interesting projects was investigating paleolake chronologies in Eastern Oregon, which included late Pleistocene shorelines adjacent to the Paisley Caves, one of the earliest human occupations in North America. Another project provided evidence of people living in the southern Willamette Valley, Oregon, as early as 9,000 yrs B.P.

Michelle Goman has spent the past 25 years studying wetland and lacustrine sedimentary archives in the United States, Mexico, and Kenya. She received her Ph.D. in Physical Geography from the University of California, Berkeley. After spending the past decade running the Quaternary Research Laboratory at the Department of Earth and Atmospheric Sciences at Cornell University, she is currently Faculty at Sonoma State University in the Department of Geography and Global Studies where she also runs the Sonoma Quaternary Lab (SQUAL). She is the Secretary of the Limnogeology Division of the Geological Society of America and was most recently the Chair of the Paleoenvironmental Change Specialty group of the Association of American Geographers. She organized the session *Reconstructing Interactions between Humans and the Natural Environment during the Holocene* at the 2010 Geological Society of America Meeting in Denver and is volume editor for the SpringerBriefs *Reconstructing Human-Landscape Interactions*.

Andrew L. Kowler has spent the past 15 years studying soils and Quaternary sediments since an undergraduate at the University of Maryland at College Park. Following his participation in 1999 National Soil Judging Competition in Tucson, Arizona, and graduation with a B.S. in Environmental Science and Policy (specialization in Land, Water, and Soil), he spent 6 years mapping soils on the Navajo and Hopi Nations in northeastern Arizona as a Soil Scientist for the USDA NRCS. In 2005, he was concurrently admitted to the University of Arizona Geosciences graduate program, and into the NSF-IGERT program in Archeological Sciences, earning his M.S. in Soil Geochemistry in 2007. As a doctoral student, he has spent the past 5 years studying paleohydrology of the southeastern Basin and Range through his work on ancient lacustrine and spring-related deposits in southern Arizona and New Mexico. In addition to involvement with geological mapping efforts within Arizona, concurrent archeological collaborations have led to involvement in the stratigraphy and geochronology of the Upper Paleolithic period in southern Tibet, the Late Paleoindian to Middle Archaic periods in the Atacama Desert of northern Chile, the Early Archaic period in coastal southern Peru, and the Early Paleoindian and Late Archaic periods in southwestern North America. Following his graduation, Andrew hopes to build on his current research and teaching program as tenure-track faculty at an academic institution.

Brian L. O'Neill is an Archaeologist whose interests include complex hunter-gatherer-fisher societies, obsidian studies, residue analysis, and the use of GIS in CRM and predictive modeling. In 1989, he received his Ph.D. in Anthropology from the University of Oregon where he is a Senior Research Associate and Staff Archeologist with the Research Division of the Museum of Natural and Cultural History. He has done field work in Missouri and Kansas, and in the Pacific Northwest in Washington and Oregon. While he has directed archeological research throughout Oregon, the focus of his work has been the interior valleys of the Willamette, Umpqua, and Rogue River drainages. He is particularly interested in documenting the early Holocene (pre-Mazama) in the western Cascades. He has authored and edited volumes in the University of Oregon Anthropological Papers, the Bureau of Land Management Cultural Resource Series, and has contributed articles to the Association of Oregon Archaeologists Occasional Papers Series and Journal of Northwest Anthropology. He has also written on Native rock art in Kansas.

Donald E. Sabol a Geologist, is an Associate Research Professor at the Desert Research Institute. With more than 25 years of experience in remote sensing research, much of his efforts focus on using remote sensing for studying spatial and temporal surface changes caused by natural and man-induced processes and spectral detection of sub-pixel targets. These have found research applications within numerous agencies, including the U.S. Army, U.S Air Force, the Department of Energy, and NASA. Before coming to DRI, he was a part of the University of Washington team for evaluation of the TES (temperature/emissivity separation algorithm) for converting ASTER thermal data to level 2 products. His focus was on validation of the image data over validation sites located in Hawaii, California, and Nevada. He holds a Ph.D. in Geological Sciences from the University of Washington and is a registered geologist in the state of Washington.

Introduction

Michelle Goman

Reconstructing Human-Landscape Interactions: *Reconstructing the Natural and Anthropogenic Landscape* is the second volume of papers published from the special session of the Geological Society of America (2010) annual meeting in Denver, Colorado, entitled *Reconstructing Interactions between Humans and the Natural Environment during the Holocene*. This oral and poster session featured cutting edge research in Geoarcheology, Geomorphology, Paleoclimatology and Paleoecology, with presentations emphasizing the complex interactions between natural environmental forces on landscape development and the increasing impact of humans on the natural environment.

This volume highlights research from the American Southwest and Northwest across late Pleistocene to late Holocene timescales. The first two chapters detail geomorphic studies from very different environmental settings. Chapter 1 by Andrew L. Kowler examines paleoshoreline evidence and inferred climate change from the Basin and Range region of Arizona and New Mexico; it primarily concentrates on late Pleistocene timescales. In Chap. 2, Dorothy E. Freidel and Brian L. O'Neill take the reader to the Pacific Northwest and discuss Early to Mid-Holocene fluvial terrace formation with implications for not only climate change but also landscape disruption from the aftermath of the Mount Mazama eruption. The final chapter by Paul E. Buck and Donald E. Sabol return the reader to Arizona. Buck and Sable use remote sensing and archeological data to develop a model to identify the optimal location of Anasazi maize fields a millennium ago. While the first two chapters focus upon geomorphic evidence for natural landscape change and the final chapter identifies the anthropogenic landscape, all three chapters are linked by their respective findings implications for Geoarcheology.

M. Goman (✉)
Department of Geography and Global Studies, Sonoma State University,
1801 East Cotati Avenue, Rohnert Park, CA 94928, USA
e-mail: goman@sonoma.edu

Introduction

Chapter 1
Lake Shoreline Evidence of Hydrologic Conditions in the Southern Basin and Range Province During the Late Pleistocene and Early Holocene: Paleoclimatic and Archaeological Implications

Andrew L. Kowler

Abstract Remnant shoreline features in hydrographically closed basins of the southern Basin and Range mark the highest stages of permanent lakes which waxed and waned during the last glacial cycle. As geomorphic footprints of these ancient lakes, fossil shorelines constitute an invaluable but largely untapped source of paleohydrologic information, with implications for the paleoclimatology and early archaeology of the region. Based on the extremely limited shoreline evidence on hand, it is known that lake highstands occurred in several basins prior to 14 Ka cal BP, while the timing of most is not confidently constrained. This suggests that Clovis (late Bølling-Allerød) and Folsom (Younger Dryas) times were generally more arid than previous periods, but does not provide specific information about conditions associated with these early occupations. Further development of shoreline records across the region is requisite for understanding the role of hydroclimatic variability in the geomorphic evolution of landscapes through the Pleistocene-Holocene transition. In turn, such knowledge will leave archaeologists better equipped to interpret spatio-temporal patterns in the cultural record with respect to the land use and settlement strategies of ancient foragers, who depended upon dwindling supplies of surface water and accompanying ecological resources.

A. L. Kowler (✉)
Department of Geosciences, University of Arizona, Tucson, AZ, USA
e-mail: kowler@email.arizona.edu

M. Goman, *Human Environment Interactions – Volume 2*,
SpringerBriefs in Earth System Sciences, DOI: 10.1007/978-3-642-36880-6_1,

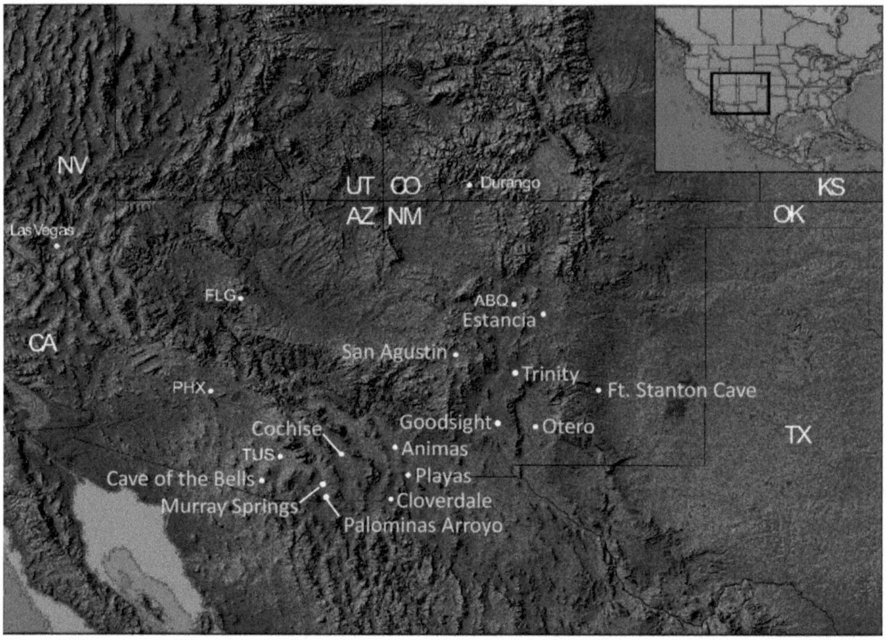

Fig. 1.1 Shaded relief map showing several of the paleolake basins in southeastern AZ and southern NM, including those referred to in Table 1.1, Figs. 1.2 and 1.8, or the Synopsis section: Paleolakes Animas (Animas Basin, Lower Animas sub-basin), Cloverdale (Animas Basin, Cloverdale sub-basin), Cochise (Willcox Basin), Estancia (Estancia Basin), Goodsight (Lake Goodsight Basin), Otero (Tularosa Basin, Lake Otero sub-basin), Playas (Playas Basin, Upper Playas sub-basin), San Agustin (San Agustin Basin), and Trinity (Lake Trinity Basin)

1.1 Introduction

1.1.1 Impetus

Late Pleistocene age remnant shoreline features mark the highest stages of permanent lakes that once covered the floors of undrained basins in the southeastern Basin and Range in Arizona (AZ) and New Mexico (NM) (Fig. 1.1) (Meinzer and Kelton 1913; Meinzer 1922; Allen 2005). As the geomorphic "footprints" of ancient lakes, paleoshoreline landforms have implications for paleoclimatology of the Pleistocene-Holocene transition (defined here as 14–10 calendar Ka) and early archaeology of the region. Lake-level chronologies extracted from fossil shorelines and coeval lacustrine sediments have the potential to shed light on the timing and relative extents of successive lake expansions, revealing spatio-temporal patterns of hydrologic change during early occupations of the region. The geographic scope of this summary is limited to hydrographically closed basins in southern NM and southeastern AZ, where my current research is focused. This review, supplemented by previously unpublished data, provides a framework for further development of paleohydrologic reconstructions for the region. When better developed, shoreline records will provide a firm hydroclimatic and geomorphic context within which to reevaluate trends in the archaeological record.

1.2 Background

1.2.1 Regional Overview

The southern Basin and Range extends southward from the Colorado Plateau and Great Basin to the USA-Mexico boundary, and westward from the Pecos River Valley to the lower Colorado River Valley, and is characterized by dozens of discontinuous block-fault mountain ranges that bound and alternate with depositional basins. The province is comprised of six physiographically distinct sections: (1) lower Colorado River Valley, (2) Sonoran Desert, (3) Mexican Highland, (4) Transition Zone-Datil-Mogollon, (5) Rio Grande Rift, and (6) Sacramento. Many of the smaller drainages within the study area have been hydrographically closed since the mid-Pleistocene or earlier (Connell et al. 2005), accounting for the numerous pluvial lakes that formed there during the late Quaternary period (Hawley 1993). Higher summits within the region range in elevation from 900–2,500 m, while basin floor elevations range between 1,200 and >2,100 m. Two wet seasons punctuate the otherwise arid climate of this region, resulting in a semi-arid climate. The North American summer monsoon, a regional-scale atmospheric circulation pattern resulting from westward expansion of the Bermuda High (Adams and Comrie 1997), produces convective storms that account for over half of mean annual precipitation. The remainder is delivered during the cooler months by Westerly-driven Pacific frontal storms. Elevations within the study area range considerably, from ~50 m in the Gila River Valley in AZ to >3,200 m in the Sandia Mountains of NM, causing significant spatial variability in mean annual temperature and precipitation.

1.2.2 Lake Dimensions and Their Hydrologic Significance

Paleolakes formed under conditions of positive hydrologic balance, whereby precipitation:evaporation (P:E) > 1:1. Additionally, a hydrographically closed basin by definition has no drainage outlet, allowing the surface area of a lake to adjust to the steady-state hydrologic balance of the basin. In contrast, lake depth and volume are indirectly related to P:E and primarily controlled by non-climatic factors such as basin geometry. The following mass-balance equation describes the general relationship between lake dimensions and climatic conditions for closed basin lakes with negligible groundwater inputs:

$$A_L = V_Q / (E_L - P_L)$$

in which A_L = lake area; V_Q = volume of stream flow discharge, including runoff generated by precipitation over land; E_L = lake evaporation rate; and P_L = precipitation falling directly onto the lake (Benson and Paillet 1989).

Although A_L responds to change in climatic conditions, basin area (A_B) also exerts a large influence on A_L (Fig. 1.2a). Thus, spatial variability in climate cannot be assessed via comparison of A_L values from coeval shorelines in different basins. Instead, one must first account for the influence of basin size on lake area by normalizing A_L with respect to A_B to obtain fractional lake area ($A_F = A_L/A_B$). Compared to A_L, the greater climatic sensitivity of A_F is apparent when plotted against elevation (Allen 2005) or modern P:E, which in turn are strongly related to one another (Fig. 1.2b). Under the assumption that the highest, well-preserved paleoshoreline in each basin formed under the most humid conditions of the last glacial cycle, corresponding A_F values should reflect the relative hydrologic "potential" of each basin. Indeed, there is a stronger correlation between P:E and paleo-A_F values from closed basins in

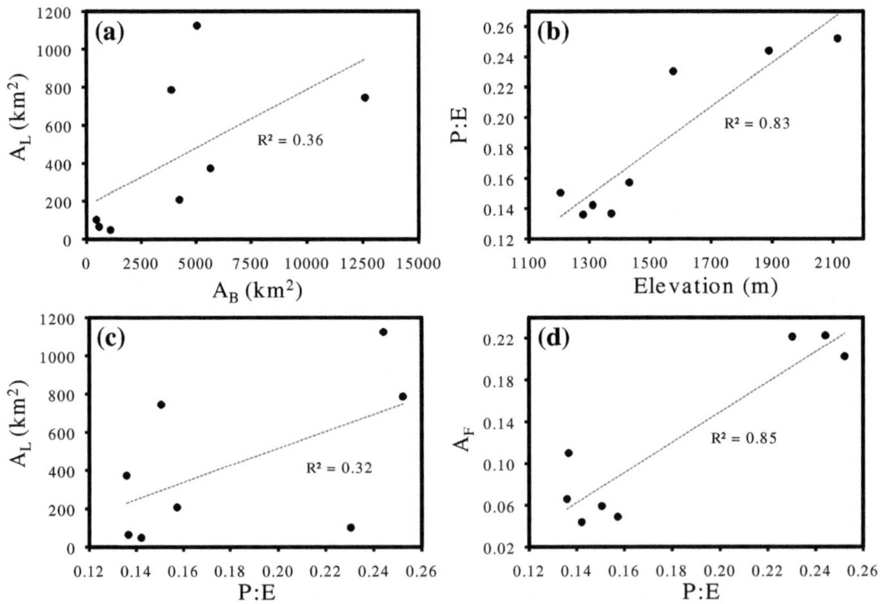

Fig. 1.2 Plots of climatic and physiographic parameters for select paleolake basins in NM: **a** paleolake area (A_L) versus basin area (A_B), **b** modern P:E versus basin floor elevation, **c** paleolake area versus modern P:E, and **d** fractional paleolake area (A_F) vs. modern P:E

NM, calculated from data reported in Allen (2005) (Fig. 1.2c, d, respectively). Mifflin and Wheat (1979) presented a more sophisticated analysis of high paleoshorelines in Nevada, serving as the basis for development of the Pluvial Hydrologic Index. Through regression analysis, they found that latitude and altitude explain 75 % of Pluvial Hydrologic Index variance among paleolake basins in Nevada, whereby 1° latitude and 500 feet of altitude correspond to an equivalent change (Mifflin and Wheat 1979).

Mifflin and Wheat (1979) found that precipitation, temperature, lake evaporation, and runoff vary exponentially in relation to latitude and altitude, and that the response of lake area to changes in precipitation is non-linear. This owes to the fact that runoff accounts for most of the precipitation reaching a lake and is itself exponentially related to precipitation. Moreover, spatial and seasonal variations of evapo-transpiration within a basin exert significant influence on runoff. In addition to temperature and relative humidity, factors that influence evapo-transpiration over land include the type and density of vegetation cover (causing increased shading and concomitant reduction in soil temperature), plant respiration (positively correlated with transpiration), and aspects of precipitation such as its style/intensity, seasonality, and the fraction delivered frozen. Although lake evaporation accounts for a much smaller fraction of total evaporation than does evapo-transpiration over land, it is nonetheless an important component. Factors that influence it include the temperature of air and water at the lake-atmosphere interface, wind velocity, relative humidity, winter ice cover, and the surface area-to-volume ratio (a function of basin geometry) of a given lake. In sum, isolating a paleoclimatic signal from shorelines cannot be accomplished solely on the basis of meteorological records.

While shoreline chronologies provide temporally discontinuous records of lake area, with multi-decadal to millennial temporal resolution, they do not often contain evidence for intervening arid intervals when lacustrine sedimentation may have either ceased or occurred only in low-lying areas. In contrast, proxy records extracted from sediment cores can provide more continuous records of lake-level change, with annual to centennial resolution. Unfortunately, such records are often compromised by the geochronologic uncertainties associated with ^{14}C dating of lake-authigenic materials, including organic matter and the remains of aquatic organisms (Deevey et al. 1954; Broecker and Walton 1959; Benson et al. 1990). As a result, sediment core records frequently rely on age-depth models for comparison to high-resolution paleoclimate time-series from other regions of the world (e.g. records from speleothems, marine sediments, glacial ice, and corals). Paleoshorelines are thus unique in that they have the potential to provide accurate and unambiguous records of lake area, thus offering a means by which to assess the timing and relative magnitudes of sustained P:E increases—information needed to calibrate age models and transfer functions that underpin proxy records. Pairing of shoreline and core records might provide enough detail to assess past changes in meso-scale atmospheric circulation and shed light on the causes of regional aridification during the Pleistocene-Holocene transition.

1.2.3 Limitations of Shoreline Records

The limitations of employing hydrologic balance modeling for extracting paleoclimatic information from relict shoreline features were first examined in the Estancia Basin in central NM by Leopold (1951) and Antevs (1954). Uncertainty in the accuracy of paleolake elevation as inferred from paleoshorelines limits the reliability of paleo-A_L reconstructions. Mifflin and Wheat (1979) suggested that underestimation of A_L, stemming from incorrect assumptions about the relation of paleoshorelines to lake-level, might account for much of the unexplained 25 % of Pluvial Hydrologic Index variance. Other limitations of shoreline records stem from the fact that paleoshorelines (1) form only in response to highstands of sufficient duration and/or degree of wave activity, (2) were subject to erosion immediately following deposition, (3) can be obscured by burial, (4) are often composite landforms made up of stratigraphic members exhibiting variability in thickness, spatial extent, and degree of preservation, and (5) do not always contain a sufficient quantity of dateable material.

Even in the event that a paleoshoreline is well-dated and allows for unequivocal determination of paleolake level, local paleoclimatic variations caused by rain shadows, and the effect of basin geometry, geology, and ecology on evapo-transpiration, are among several factors precluding the use of A_L values for regional climatic reconstructions. In addition, it is difficult to test the assumption that topographically closed basins were geohydrologically closed in the past. Mifflin and Wheat (1979) warned that even if groundwater leakage is currently negligible, increased storage could have resulted in an increase in, and/or reversal of, inter-basin flow (Mifflin and Wheat 1979). According to Hawley (1993), most or all closed basins in southern NM are not geohydrologically closed, but are instead characterized by deep groundwater flow into and/or out of the basin. Thus, most paleohydrologic studies that do not take groundwater-to-surface water interactions into account (Meinzer 1922; Leopold 1951; Antevs 1954; Langbein 1961; Brackenridge 1978; Smith and Anderson 1982; Galloway 1983) do not fully address the hydroclimatic significance of fossil shorelines (Hawley 1993).

1.2.4 Lake-Level Reconstruction from Shoreline Landforms

Physiography of the coastline strongly influences the geomorphic processes that act on any given segment, producing an array of landform types with varying relation to lake-level. Shoreline features include both constructional and erosional landforms. Because beach deposits were emplaced by waves, they are reliable lake-level indicators. However, because not all constructional shoreline features have a beach origin, lake-level reconstruction from other landforms requires careful consideration of geomorphic, stratigraphic, and sedimentologic evidence (Powers 1939; Fleischauer and Stone 1982).

In Paleolakes San Agustin and Animas, the irregular shape of the coastline caused embayments to form, in turn influencing the development of shoreline landforms. Landforms that typically developed in these paleo-embayments (Fig. 1.3) include (1) bay mouth bar, which spans the entrance of an embayment and separates it from the main body of the lake, (2) bayside beach, which lines the interior coastline of an embayment, and (3) spit, formed from the progradation of coarse sediments from headlands into shallow areas protected by them via longshore transport (Powers 1939). Bay mouth bars may have formed as merged spits, as suggested by their lakeward-descending elevation (Fleischauer and Stone 1982). Following its emplacement, a shoreline deposit can serve as a substrate for subsequent aggradation, such that a polygenetic origin must always be considered.

Not only is interpretation of landform type difficult, but reconstruction of ancient shoreline elevations from them can be inexact. Powers (1939) observed that the vertical differences between shoreline features marking different stages of Paleolake San Agustin are of the same order as internal variations within the landforms themselves (Fig. 1.3). For instance, one probable beach ridge of Paleolake San Agustin exhibits lateral variation in elevation of up to 13 feet. In addition, although spits are similar in composition to beaches, their snouts can be several meters lower than the associated shoreline. Fleischauer and Stone (1982) made similar observations with respect to shoreline features of Paleolake Animas.

Further complicating matters is the possibility that some of this variability is the product of the lake's regression from or transgression to an adjacent contour. Powers (1939) posited that such variations observed in the San Agustin basin might have arisen from varying degrees of exposure to storm waves or differing distance from the sediment source. Another possibility is that the higher parts of a shoreline ridge are comprised of storm beaches, whereas the lower parts originated as subaqueous bars. Fleischauer and Stone (1982) suggested that the constructional landforms associated with Paleolake Animas were polygenetic, concluding that

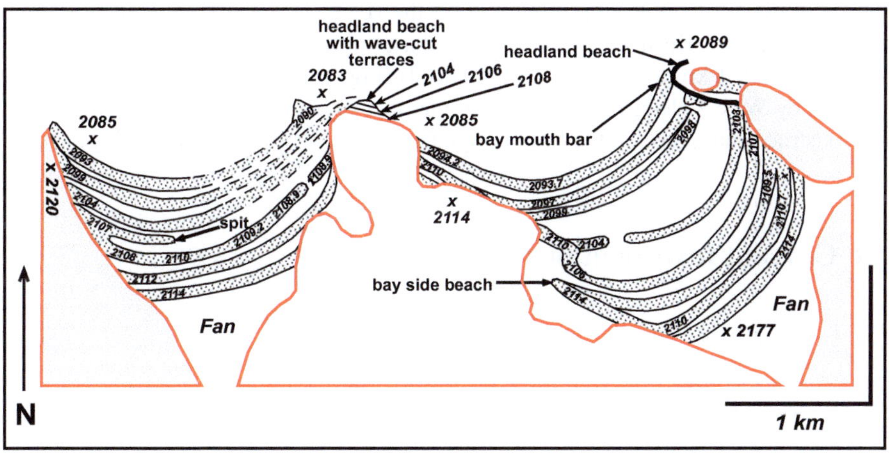

Fig. 1.3 Detail of paleoshoreline landforms in the southeastern portion of the San Agustin Basin, exhibiting a variety of morphologies typically associated with embayments (after Powers 1939). Elevations are expressed in meters above mean sea level

estimation of paleolake level requires the combined use of stratigraphy and shoreline morphology. Employing a quantitative approach, Powers (1939) constructed an elevation histogram for 201 beaches, bars, spits, and wave-cut benches, identifying five distinct levels between 2,115 and 2,084 m and observing that shoreline landforms at 2,115 m occur around the entire circumference of the basin floor.

1.3 Investigations in the Willcox Basin

1.3.1 Previous Investigations

Past investigations that provide stratigraphic and chronologic context for my current work in Willcox Basin, southeastern AZ are described in this section. Several anthropogenic exposures of bedded sandstones and conglomerates offer clues about the lacustrine origin of the composite gravel ridge found along the western margin of Willcox Playa between 1,275 and 1,271 m, leading Meinzer and Kelton (1913) to propose the existence of a paleolake with corresponding dimensions during the last glacial cycle. Robinson (1965) and Schreiber (1978) reported the results of detailed sedimentologic investigations along this ridge, corroborating and expanding upon the findings of Meinzer and Kelton (1913), but providing no age control for the beach gravels. Long (1966) defined two distinct late Pleistocene lake cycles based primarily on ^{14}C dating of the "Willcox White Marl" and the carbonate nodules that he collected from its denuded surface above the eastern periphery of Willcox Playa. During Long's (1966) first proposed lake cycle from >33–16 Ka, the lake would have reached >1,290 m, while only reaching the elevation of the beach ridge (~1,274 m) between 13.5 and 12 Ka. Haynes et al. (1987) described the stratigraphy of beach and nearshore deposits from an exposure along the ridge (Locality A, Fig. 1.4), obtaining a date of 15.6 Ka from charcoal in the basal portion of the section, in Unit B3. Approximately 1 km to the south at Locality B (Fig. 1.4), five ages on charcoal were reported by Waters (1989) from the lower member of his Unit B, constraining an earlier constructional phase of the beach ridge from ≥16.9–16.4 Ka or later (Fig. 1.5).

1.4 Current Investigations

1.4.1 Methods

1.4.1.1 Mapping and Stratigraphy

Detailed mapping of shorelines in the Willcox Basin (results forthcoming) enabled relocation of stratigraphic localities reported in Haynes et al. (1987) and Waters (1989) (Fig. 1.4). The Locality A transect approximates the location and orientation

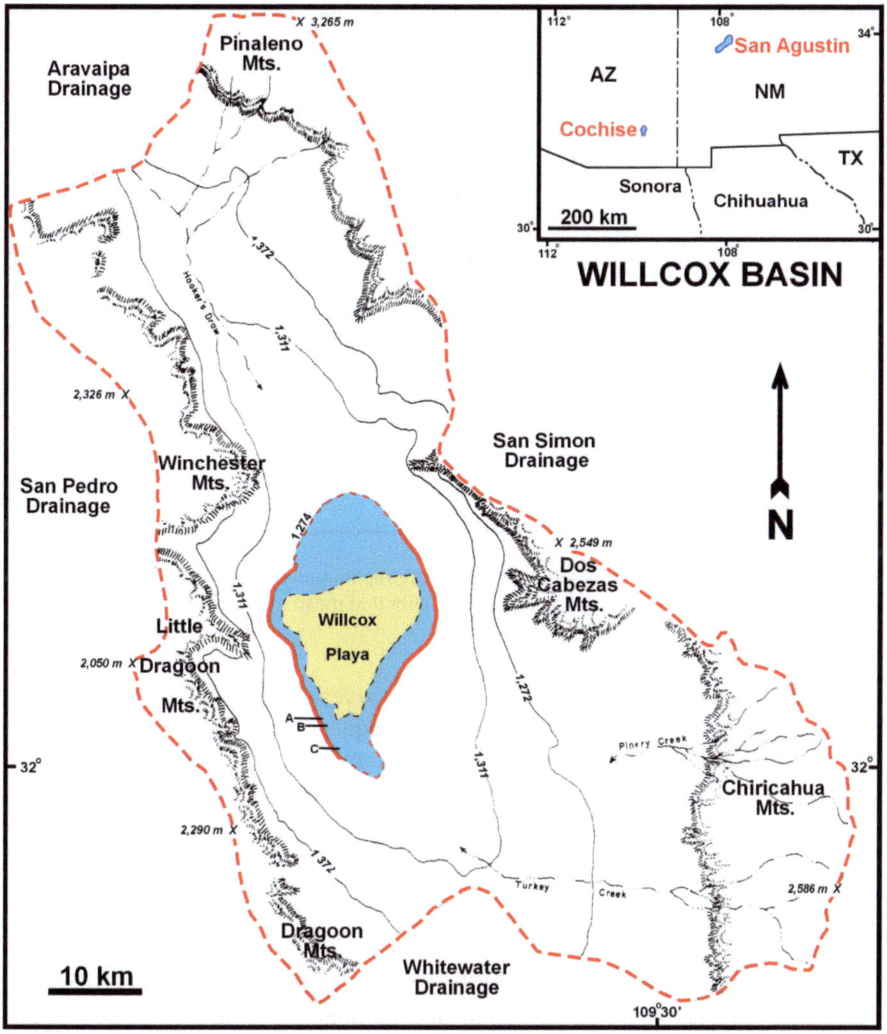

Fig. 1.4 Diagram of Willcox Basin, with inset regional map. *Dashed red line* demarcates the hydrologic divide with adjacent basins. The yellowish-tan area represents the area presently covered by Willcox Playa, while the *blue area* represents the surface area of Paleolake Cochise when it constructed the beach deposits associated with the 1,271–1,275 m contour swath. *Bold black lines* intersecting with the western paleoshoreline of Paleolake Cochise (*solid red line*) demarcate the stratigraphic localities depicted in Figs. 1.5 and 1.6

of the stratigraphic profile reported in Haynes et al. (1987), whereas localities B and C correspond to the Cameron and Grim localities, respectively, of Waters (1989). Elevations along the Locality C transect (Fig. 1.6) were estimated on the basis of an abney level, as well as topographic maps and horizontal coordinates from a

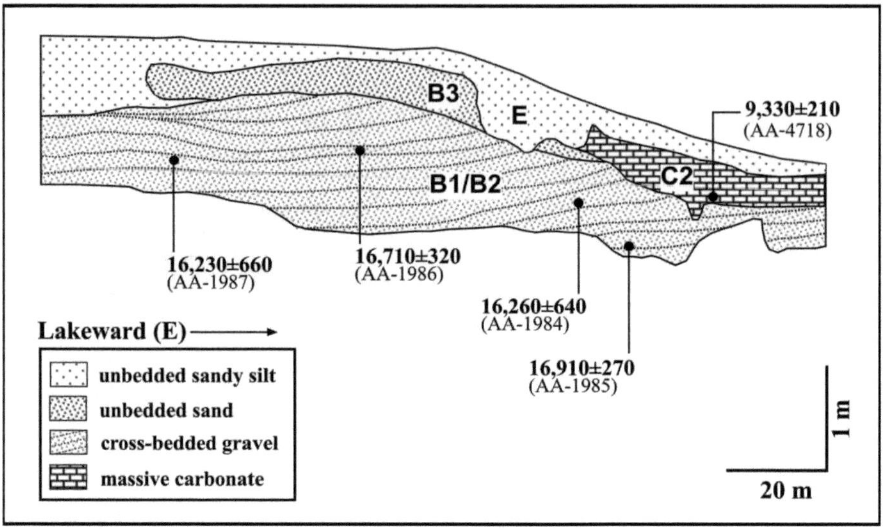

Fig. 1.5 Stratigraphic fence diagram depicting stratigraphic units comprising the beach ridge at Locality B in Willcox Basin (shown in Fig. 1.4). Calibrated mean aages are expressed in calender year ranges (reported in Table 1.2)

hand-held GPS unit, although a more accurate survey is planned. Extensive trenching along the Locality C transect exposed a sequence of lacustrine and wetland deposits to a maximum depth of 4 m. Stratigraphic profiling (Fig. 1.6) and ^{14}C analyses (Table 1.2) were conducted between the spring of 2009 and the autumn of 2012.

1.4.1.2 Radiocarbon Dating

Gastropod shells were rinsed in distilled water and inspected under binocular microscope for secondary cementation and taphonomic evidence of reworking, in addition to their degree of preservation. Depending upon resilience of the shell material, specimens were briefly sonicated (2–5 s) up to 3 times, prior to a final rinse with distilled water. Shells were then crushed and their interior and exterior surfaces inspected for adhering detritus. Visibly impure fragments were either discarded or scraped clean with a stainless steel probe. Both shell and sedimentary (non-biogenic) carbonate specimens were immersed in H_2O_2 for complete removal of organic matter. For ^{14}C dating of carbonates, each sample underwent in vacuo hydrolysis in 100 % H_3PO_4, followed by cryogenic extraction of CO_2, at the University of Arizona AMS radiocarbon facility. Impurities (H_2O, SO_x, NO_x, and halides) were removed with Cu and Ag traps at ~600 °C, and the purified CO_2 was divided into two portions. One aliquot was converted to graphite by catalytic reduction of CO and submitted for ^{14}C analysis, while the other provided $\delta^{13}C$, needed to correct the measured ^{14}C activity for isotopic fractionation. For

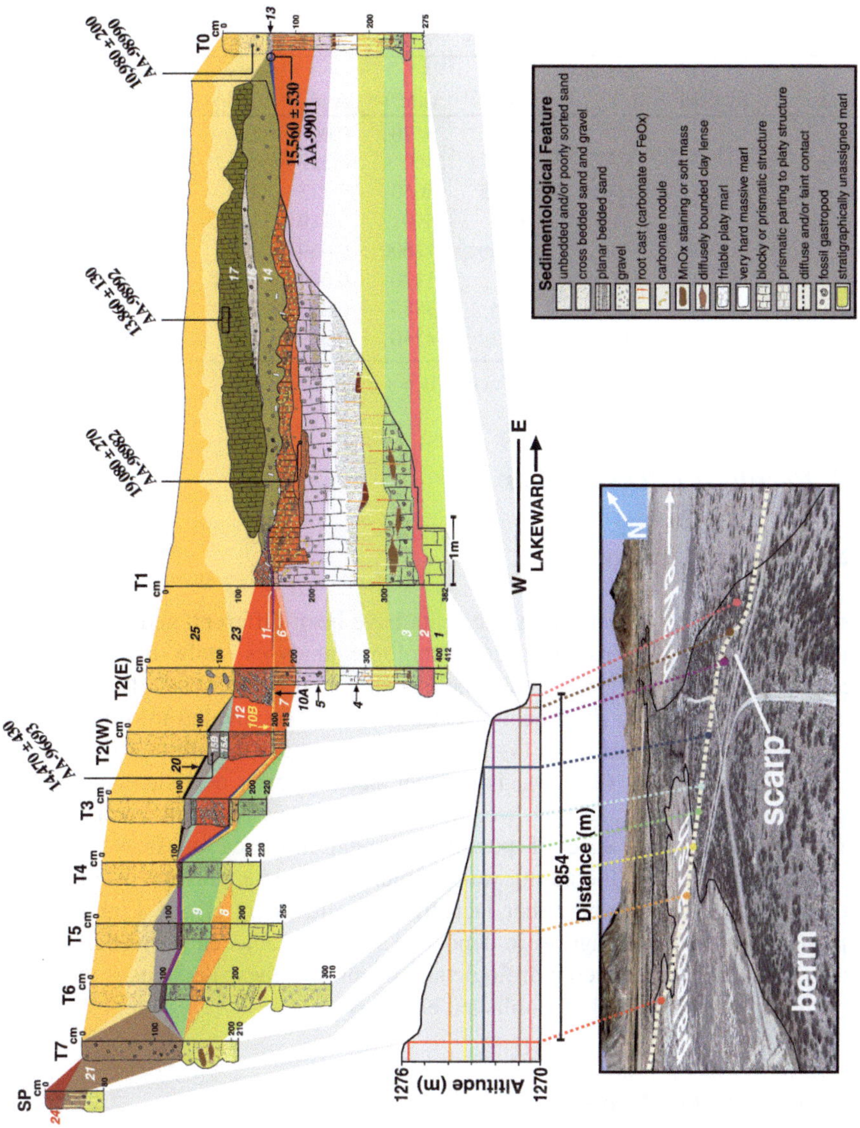

Fig. 1.6 Stratigraphic fence diagram depicting shoreline deposits associated with the beach deposit exposed in skip trenches excavated at Locality C in Willcox Basin (shown in Fig. 1.4). Radiocarbon ages are expressed in calender year ranges (reported in Table 1.2)

conversion of ^{14}C dates (^{14}C yr BP) to calendar ages, CALIB 6.0.1 (IntCal09) (Stuiver and Reimer 1993; Reimer et al. 2009) was used to generate a 2-sigma probability range and corresponding median value (expressed as 10^3 cal yr BP, denoted as Ka) (Table 1.2).

Table 1.1 Physical and climatic parameters for closed basins and associated paleolakes in southern NM (after Allen 2005)

Paleolake	Symbol	Elevation (m)	A_L (km^2)	A_B (km^2)	A_F	Precip (cm/yr)	Evap (cm/yr)	P:E
Animas	A	1279	374	5670	0.066	25	184	0.136
Cloverdale	Cl	1576	102	460	0.222	41	178	0.230
Estancia	Es	1890	1125	5050	0.223	31	127	0.244
Goodsight	G	1372	65	590	0.110	25	183	0.137
Otero	O	1204	745	12600	0.059	25	166	0.151
Playas	Pl	1311	49	1120	0.044	27	190	0.142
San Agustin	SA	2115	786	3880	0.203	29	115	0.252
Trinity	T	1431	207	4240	0.049	25	159	0.157

Information for Willcox Basin/Paleolake Cochise is not included

1.4.2 Results and Discussion

1.4.2.1 Lithostratigraphy

Thus far, twenty-five lithostratigraphic units have been differentiated on the basis of field-based observations and geochronologic analyses, only twenty-two of which are exposed in sections along the Locality C transect (Figs. 1.4 and 1.6). The sequence revealed in the Locality C trenches can be divided into three basic parts: lower clay and marl units (1–9), middle coarse-grained units (10–20), and upper fine-grained units (21–25). The lower package was deposited under low energy conditions associated with wetland and littoral environments, and contains secondary features indicative of post-depositional modification from bioturbation, pedogenesis, and a fluctuating water table. The middle package is comprised mostly of sands and gravels, either lacustrine or alluvial in origin. Finally, units comprising the upper package were deposited within or adjecent to wetlands. The depositional history and paleohydrologic significance of these deposits is the focus of my current work and will be published elsewhere.

1.4.2.2 Radiocarbon Dating

Carbonate specimens from Locality 3 were selected for ^{14}C dating, including three gastropods (Units 7, 15B, and 23) and bulk sedimentary carbonate from Units 13 and 17, resulting in five new dates between 19 and 11 Ka (Table 1.2; Fig. 1.6). Shells were selected for dating on the basis of their identification as Succineidae, a family of pulmonate (lung-breathing) semi-aquatic snails known to yield reliable ^{14}C ages, with the exception of very few species. This is because most Succineidae species utilize decaying plant matter as their primary source of carbon for shell building, such that the initial ^{14}C activity of the shell is in equilibrium with atmospheric ^{14}C (Pigati et al. 2004, 2010). Further, ^{14}C contamination through open

Table 1.2 Radiocarbon dates corresponding to stratigraphic fence diagrams (Figs. 1.5 and 1.6) along with one reported in Haynes et al. (1987).

					Cal yr BP (2 σ)						
Lab #	Material	$\delta^{13}C$ (VPDB)	^{14}C yr BP	±	Midpoint	±	Median	Unit	Locality	Figure	Reference
AA-4718	Humate	23.4	8,390	100	9,310	170	9,350	B1/B2	B	5	Waters (1989)
AA-98990	Shell	−3	9,630	50	10,980	200	10,960	23	C	6	This study
AA-98992	Sediment	−2.4	11,990	50	13,860	130	13,850	17	C	6	This study
AA-96693	Shell	−3.2	12,350	60	14,470	430	14,380	15B	C	6	This study
AA-99011	Cement	0.0	12,910	60	15,560	530	15,430	13	C	6	This study
AA-866	Charcoal	−22	12,970	480	15,440	440	15,570	B3	A	–	Haynes et al. (1987)
AA-1987	Charcoal	−22	13,380	150	16,230	660	16,460	B1/B2	B	5	Waters (1989)
AA-1984	Charcoal	−22	13,400	130	16,260	640	16,530	B1/B2	B	5	Waters (1989)
AA-1986	Charcoal	−22	13,590	120	16,710	320	16,760	B1/B2	B	5	Waters (1989)
AA-1985	Charcoal	−22	13,750	120	16,910	270	16,890	B1/B2	B	5	Waters (1989)
AA-98982	Shell	−2.5	15,860	60	19,080	270	19,080	7	C	6	This study

Radiocarbon means (with 1sd error), as well as calibrated age ranges and corresoponding median probabilities—generated using Calib 6.0—are reported.

system behavior of carbon in gastropod shells is extremely improbable within the reported age range (Rech et al. 2011 and sources therein). However, dates from sedimentary carbonates are considered as minimum ages; not only are these materials subject to open-system behavior, but their mineralogical composition prevents detection of secondary calcite.

A snail from Unit 7 dating to 19.1 Ka represents either a highstand or wetland expansion along the 1,271 m scarp during the terminal part of the Last Glacial Maximum (LGM). Dates on charcoal originally reported in Waters (1989) reflect the timing of one or more constructional phases of the beach ridge between ~17 and ~16 Ka. A date of 14.4 Ka on pedogenic carbonate from Unit 15B supports this, and a date of 15.4 Ka on bulk carbonate from Unit 10A is in close agreement with the charcoal age originally reported in Haynes et al. (1987). Bulk carbonate from the top of Unit 17—an unconsolidated, calcareous silty fine sand—yielded an age of 13.9 Ka, providing a minimum age for cessation of Unit 17 deposition and constraining the timing of the last documented regression of Paleolake Cochise. Finally, a date of 11.0 Ka from a snail in Unit 23 provides evidence for spring activity along the late Pleistocene shoreline.

1.5 Synopses of Southern Basin and Range Lake-Level Records

Paleolakes that formed in hydrographically closed basins in southeastern AZ and southern NM during the last glacial cycle include Paleolakes (1) Cochise (Willcox Basin), (2) Animas (Animas Basin, Lower Animas sub-basin), (3) Playas (Playas Basin, Lower Playas sub-basin), (4) Estancia (Estancia Basin), (5) San Agustin (San Agustin Basin), (6) Cloverdale (Animas Basin, Cloverdale sub-basin), (7) Otero (Tularosa Basin, Lake Otero sub-basin), (8) Trinity (Lake Trinity Basin), (9) Encino (Encino Basin), (10) Pinos Wells (Pinos Wells Basin), and (11) Sacramento (Sacramento Basin) (Schwennesen 1918; Meinzer 1922; Hawley 1993; Allen 2005). Following are synopses of lake-level reconstructions for Paleolakes Estancia, San Agustin, Cloverdale, and Cochise, collectively revealing that several highstands occurred throughout the region during the LGM through terminal Pleistocene interval—while also exposing a large gap in our knowledge about the status of lake levels during the period of interest (14–10 Ka). The following summaries may differ from published interpretations, as they reflect my own assessment of published ^{14}C data in light of accompanying contextual information.

1.5.1 Paleolake Estancia

While Estancia Basin is topographically closed, there is evidence of groundwater leakage into the Tularosa Basin through a solution-enlarged fracture zone along the southern divide, and it is likely that the impact of this leakage became proportionately larger as the lake deepened (Hawley 1993). As a result, the highest shoreline (1,890 m) may reflect a hydrologic threshold that prevented the lake from rising further, limiting our ability to distinguish extremely wet from moderately wet intervals.

The Paleolake Estancia record is the synthesis of several proxy data sets, including (1) sedimentological (grain-size), (2) geochemical (Sr/Ca ratios), (3) faunal (ostracoda), (4) stratigraphic and geochronologic (nearshore sediments), and (5) geomorphic (Allen and Anderson 2000; Anderson et al. 2002). Collectively, these records indicate that Paleolake Estancia experienced maximum freshening and probably reached 1,890 m on several occasions between 23 and 14 Ka. A ^{14}C date of 14.6 Ka from the top of a continuous, >15 Ka-long lakebed sequence places a maximum constraint on the timing of terminal Pleistocene desiccation. Following this, the lake either remained desiccated or experienced numerous oscillations reaching intermediate or lower paleoshoreline levels. On the basis of ^{14}C dating of nearshore sediments (Allen and Anderson 2000), it is clear that lake expansions reached a maximum elevation of 1,875 m at 13.4 and 10.4 Ka. Desiccation occurring at some point after 13.4 Ka led to truncation of the uppermost portion of the lake bed, which was subsequently redeposited into a gypsum sand sheet. Low-lying, dune-like landforms positioned between 1,860 and 1,865 m likely formed from subsequent eolian reworking of the sandsheet along the dry margin of a wetland or lowstand lake. Anderson et al. (2002) speculated that a lowstand resulted in subaqueous modification of these dunes during the Younger Dryas chronozone (12.9–11.7 Ka), and ^{14}C data reported in (Anderson et al. 2002) constrains the timing of a possible lowstand to the period between 13.4 and 11.1 Ka.

1.5.2 Paleolake San Agustin

The San Agustin Basin has yielded limited paleoenvironmental information pertaining to the period of interest; despite the numerous attempts at reconstructing its lake-level history, the lack of dated shorelines precludes determination of maximum lake elevation from existing data. This ~90 km-long basin contains three interconnected sub-basins, named Horse Spring (2,065 m), C–N (2,102 m), and White Lake (2,120 m). According to proxy-based lake-level reconstructions derived from sediment cores in combination with mapped shorelines (Weber 1994), all three sub-basins were submerged by a fresh water lake during the LGM (Markgraf et al. 1983, 1984; Phillips et al. 1992). At some point after 19 Ka the White Lake Basin dried and marsh-like conditions developed in the C–N Basin, leaving a fresh water lake only in the adjacent Horse Spring Basin. There, radiocarbon dates from Bat Cave (Holliday et al. 2006) show that the shoreline had receded to 2,105 m or lower by 13.2 Ka, after which it transitioned from fresh to alkaline and then saline conditions prior to drying up in the middle Holocene. In sum, evidence points toward irregular reduction of total lake area within the San Agustin Basin after ~15 Ka until final disappearance of the Horse Spring lake during the middle Holocene. However, a more detailed reconstruction of paleo-hydrologic changes during the Pleistocene-Holocene transition awaits dating of shoreline landforms, along with the establishment of a more complete and reliably dated sediment core record.

1.5.3 Paleolake Cloverdale

The most prominent paleoshoreline in the Upper Animas basin records a highstand of Paleolake Cloverdale during the LGM, along with three subsequent lake stands during the middle and late Holocene (Krider 1998). As yet, there is no evidence of high-stands during the Pleistocene-Holocene transition, although this could be the result of a limited effort toward stratigraphic examination and dating of lacustrine features.

1.5.4 Paleolake Cochise

The results of stratigraphic and geochronologic investigation of the prominent shoreline berm along the western margin of the playa (1,275–1,271 m) reported here and elsewhere (Haynes et al. 1987; Waters 1989) suggest that the beach was constructed by several highstands between ~17 and 14 Ka—corresponding closely to the hiatus in Long's (1966) record. A possible explanation for this is that Long's (1966) carbonates formed in association with the capillary fringe or extensive spring discharge, and not a lake. It is therefore reasonable to speculate that Long's (1966) "pluvial" periods actually reflect a hydrologic state under which seeps and artesian springs were active, supplying streams that in turn fed a valley bottom wetland. Perhaps the hydrologic state of Willcox Basin alternated between (1) extensive spring discharge with no lake or a lowstand, versus (2) curtailed groundwater discharge with a highstand, each state resulting from a distinct configuration of hydrologic and climatic conditions. In conclusion, the period of interest (14–10 Ka) can be tentatively divided into four parts: (1) a highstand interval ending by or shortly after 14 Ka, (2) a dry interval from 14–13.5 Ka, (3) a period characterized by intermittent spring activity from 13.5–12 Ka, and (4) a relatively dry interval from 12–10 Ka that was interrupted by brief wet intervals.

1.6 Paleohydrologic Records from the Southern Basin and Range and Their Climatic Implications

While sediment cores obtained from the San Agustin and Estancia Basins have provided information used to construct detailed paleohydrologic records, geochronologic uncertainties and a paucity of dated shorelines prevent us from

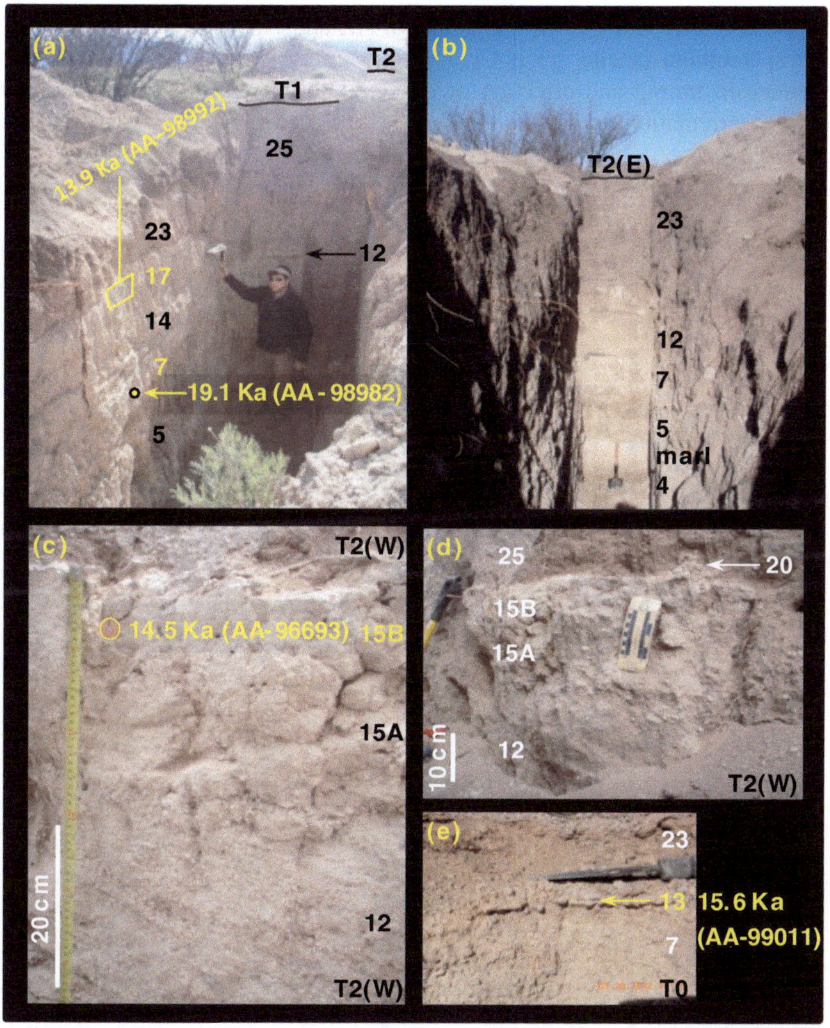

Fig. 1.7 Stratigraphy exposed in trenches along the Locality 3 transect (Fig. 1.4), showing dated units: **a** west-facing view of T1, **b** east-facing view of T2(E), **c** close-up of units 12 and 15 exposed in the southern wall of T2(W), **d** units 12, 15, and 20 exposed in the southern wall of T2(W), and **e** close-up, east-facing view of units 7, 13, and 23 in the lakeward end of T0

confidently reconstructing the timing and extent of past lake expansions in most basins (Fig. 1.1). Conversely, incomplete dating of shoreline deposits and a lack of accompanying sediment core studies limit the temporal resolution of lake-level reconstructions for Paleolakes Cochise and Cloverdale. As a result of these limitations, the late Quaternary paleohydrology of the southeastern Basin and Range remains poorly understood. Continuous $\delta^{18}O$ records from speleothems provide a qualitative framework for paleoclimatic *variability* within which to evaluate the few well-dated shoreline deposits in southern AZ and NM. Although controls on $\delta^{18}O$ in speleothem calcite are not completely understood, the remarkable correspondence in timing and direction of observed shifts suggests that changes in atmospheric circulation across the Pleistocene-Holocene boundary were regional in scale (Fig. 1.8).

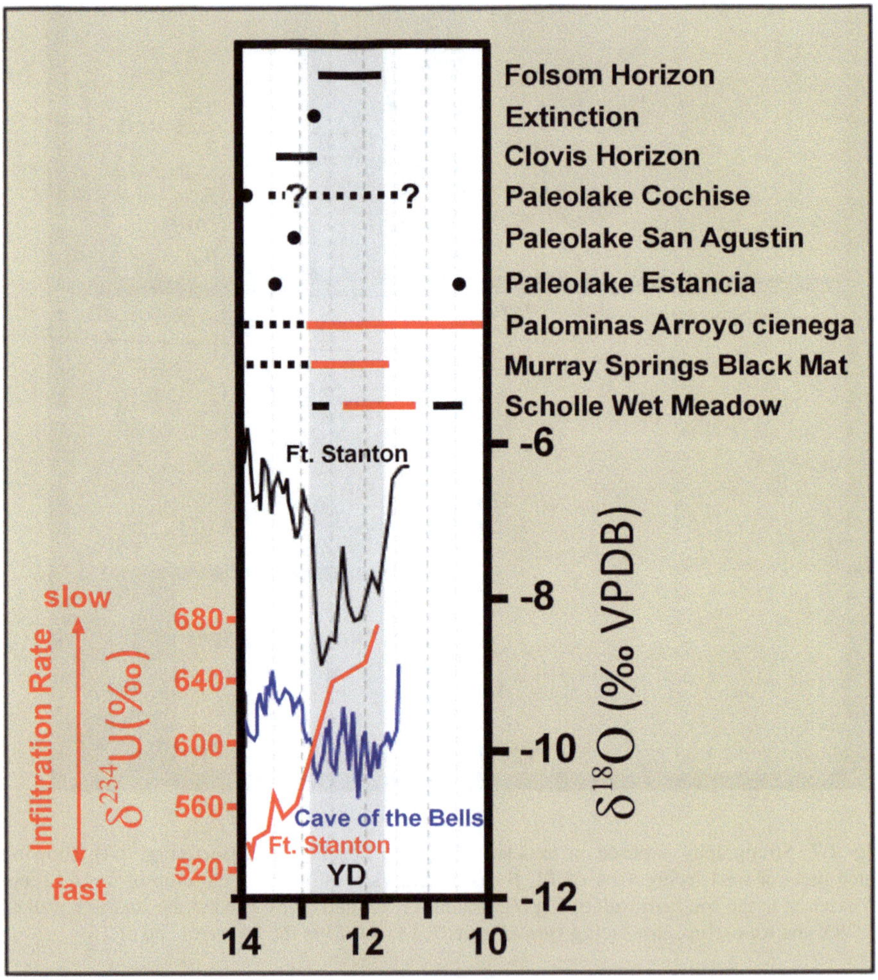

Fig. 1.8 $\delta^{18}O$ records from speleothem COB-01-02 from Cave of the Bells (*blue*) in southeastern AZ and speleothem FS-2 from Ft. Stanton Cave (*black*) in southeastern NM, and a ^{234}U record from speleothem FS-2 (*red*). Paleohydrologic records from Abo Arroyo (Scholle Wet Meadow), Murray Springs, and Palominas Arroyo are depicted with alluvial deposits in *black* and cienega deposits in *red*. Also depicted are dates from paleoshoreline deposits of Paleolakes Estancia, San Agustin, and Cochise, as well as the timing of the Clovis and Folsom cultural horizons and the extinction of remaining Rancholabrean fauna. The Younger Dryas chronozone is shaded in *grey*

Existing paleolake records suggest that lake stands prior to 15 Ka lasted for centuries or millennia and were punctuated by lowstands, while subsequent oscillations were shorter in duration and/or less expansive. Speleothem records from southeastern AZ (Wagner et al. 2010) and central NM (Polyak et al. 2004, 2012; Asmerom et al. 2010) indicate that increased summer moisture and/or an overall warming/drying trend throughout the region had begun by 14.5 Ka, coincident with the first part of the Bølling-Allerød climatic interval. This interpretation is supported by dating of the Coro Marl, a groundwater discharge carbonate deposited throughout the San Pedro Basin between >50 and 15 Ka (Haynes 2007; Pigati et al. 2009). Based on speleothem records from Carlsbad Caverns and Ft. Stanton Caves, Polyak et al. (2012) infer that an abrupt climate shift occurred ~14.6, with extremely arid conditions persisting through 12.9 Ka—at which time the onset of cooler/wetter conditions marked a brief and mild interruption of the overall warming/drying trend otherwise characterizing the Pleistocene-Holocene transition. This interpretation is based primarily on the very close correlation of ^{234}U (Fig. 1.8) and δ^{13}C data from Ft. Stanton Cave in southeastern NM, the former of which is thought to represent the relative rate of drip-water infiltration (Polyak et al. 2012). Corroborating evidence for a late glacial drying trend comes from a U-series chronology of shelfstones in Carlsbad Caverns, which indicates that pool basins dried up around 13.5 Ka (Polyak et al. 2012). It is likely that Bølling-Allerød aridity resulted in the complete disappearance of most remaining paleolakes between 15 and 13 Ka, such that any highstands would have been short-lived. Between 14 and 13 Ka, the floors of most basins had either desiccated or were covered by patchy wetlands during a period of increasing aridity interrupted by brief humid intervals. It is not clear if or when lake oscillations occurred during this interval, or if the paleohydrologic response of most basins to increased effective moisture was restricted to spring discharge. The synthesis of results from past and current studies in Willcox Basin suggests that dry conditions persisted between 14 and 13.5 Ka, after which limited spring activity persisted through 11 Ka. This scenario is in agreement with the interpretation of Polyak et al. (2012) from speleothem records in southern NM.

Speleothem records throughout the region (Polyak et al. 2004; Asmerom et al. 2010; Wagner et al. 2010; Polyak et al. 2012) indicate that an abrupt shift to cooler/wetter conditions and/or an increase in winter precipitation occurred shortly after 13 Ka, with a resumption of drying after 11.5 Ka. This long interruption of the overall warming/drying trend coincides with the Younger Dryas climatic reversal as recorded in Greenland ice cores (Alley 2000). Speleothem-based evidence from Cave of the Bells (Fig. 1.1) in the Santa Rita range (Wagner et al. 2010) corroborates the sedimentary record from Murray Springs in the San Pedro Basin (Fig. 1.1) (Haynes 2007; Pigati 2009). Between 16 and 13 Ka, the channel sands comprising stratum F1 originated in part from spring conduits. Shortly after 12.9 Ka, a brief but intense dry interval directly preceded deposition of the Clanton Clay (an organic-rich clay stratum known as the "Black Mat") and the time-equivalent Earp Marl between ~12.8 and 11.2 Ka. Deposition of the Earp Marl signifies a return to hydrochemical conditions similar in nature to those accounting for the Coro Marl (Haynes 2007), although isotopic evidence suggests that it was deposited locally in ponds—whereas the Coro Marl formed in the through-flowing waters of a cienega (Pigati 2009). Cores extracted from nearby Palominas Arroyo (Fig. 1.1) exposed a package of alluvial sands (clean sands with lenses of organic-stained sands) overlain by a package of intercalated alluvial and organic-rich paludal sediments. ^{14}C ages indicate that the lower unit was deposited between 12.9 and 11.3 Ka, and that deposition of the overlying fine-grained, organic-rich sediments began by ~10 Ka (Ballenger 2010).

A well-dated sequence is exposed in the upper Abo Arroyo drainage in central NM (Fig. 1.1), where silty alluvium dating from 12.8–12.4 Ka is overlain by a wet meadow deposit dating from 12.3–11.2 Ka (Hall et al. 2012). The near-absence of ostracoda and low abundance and taxonomic diversity of land snails in the silty alluvium indicate that it was deposited in a channel surrounded by relatively dry ground under conditions that were appreciably cooler than the extant climate. The wet meadow unit can be traced to a nearby spring head that is now buried and inactive. Accordingly, it contains a diverse assemblage of semi-aquatic and alpine terrestrial snails, as well as abundant and diverse ostracode fauna indicative of through-flowing water with a spring origin. However, this chronology remains provisional pending the results of current efforts to redate the deposits using alternative techniques (Haynes, personal communication).

In light of the aforementioned groundwater records, it is likely that Long's (1966) second "pluvial" period for Willcox Basin can be divided into two parts. From 13.5–13 Ka, spring activity was perhaps more spatially restricted, and less vigorous, than that which occurred from 13–12 Ka. Forthcoming results from the Willcox Basin will hopefully allow constraint of the timing and magnitude of the P:E changes that occurred during these intervals. Speleothem records indicate that a transition back to extremely warm/arid conditions began throughout the region between 11.5 and 10.5 Ka (Polyak et al. 2004; Asmerom et al. 2010; Wagner et al. 2010). At that time, it is likely that any remaining lakes and wetlands had desiccated, with the known exception of a lake in the Horse Spring sub-basin of the San Agustin Basin, that finally dried up during the middle Holocene (Markgraf et al. 1984). For the majority of the remainder of the Holocene, hydrologic conditions across the region probably resembled present conditions. Today, barren "playa" flats mark the wind-deflated surfaces of Wisconsin age lake beds in closed basins throughout the region.

Radiocarbon dating of shoreline deposits and associated nearshore sediments constrains the timing of highstands spanning the Pleistocene-Holocene transition (14–10 Ka) for paleolakes in the Willcox (13.9 Ka), San Agustin (13.2 Ka), and Estancia (10.4 Ka) Basins. Geomorphic evidence from the Estancia Basin indicates that a lowstand occurred there at some point(s) between 13.4 and 11.1 Ka, as indicated by data reported in Allen and Anderson (2000). These observations suggest that brief but significant P:E oscillations coincided with both the onset and termination of the Bølling-Allerød and Younger Dryas climatic intervals. Because preliminary shoreline chronologies reveal the timing of significant humid intervals that do not uniformly register in other paleoenvironmental records, they constitute an invaluable source of regional paleohydrologic and paleoclimatic information.

The absence of shoreline dates within the Younger Dryas chronozone is especially curious, given that nearshore deposits in several basins have yielded ages spanning the Bølling-Allerød climatic interval (14.5–13 Ka)—for which speleothem records (Polyak et al. 2004; Asmerom et al. 2010; Wagner et al. 2010; Polyak et al. 2012) and stratigraphic records (Haynes 1991, 2007) indicate a drying trend. Paleowetland deposits in southeastern AZ (Haynes 2007) and central NM (Hall et al. 2012) formed during the Younger Dryas chronozone, indicating elevated spring discharge levels throughout the region during much of this interval. Several scenarios might explain a lack of evidence for lakes during this interval, including that (1) wetter and/or colder winters allowed for increased aquifer recharge while runoff was curtailed by increased evapo-transpiration, in turn resulting from increased ecological productivity and warmer/drier summers, (2) lake stands were of extremely short duration, or (3) lakes reached lower levels during this interval than prior to 13 Ka. Under the latter two scenarios, resultant lacustrine deposits might have been too thin and/or low in elevation to form conspicuous shoreline features or withstand erosion. Alternatively, they may await discovery.

1.7 Archaeological Implications of Shoreline Chronologies

Fossil shorelines hold important clues to local and regional archaeological problems. First, well-dated paleohydrologic records obtained from geomorphic investigation of fossil shoreline and wetland deposits can inform us more directly about the significance of patterns and trends in the regional archaeological record than can high-resolution, continuous time-series extracted from speleothems. This is true from the standpoint that the geomorphic (shoreline) record provides unequivocal evidence for the presence of local surface water, unlike proxy records. Next, information concerning the geographic configuration and age of fossil beach and associated nearshore deposits will facilitate ongoing efforts to locate buried archaeological sites, which could provide much needed temporal and paleoenvironmental context for understanding human-landscape interactions.

1.7.1 Local

Thus far, little systematic research on the physical relationship between site location and paleoshorelines has been conducted in the southeastern Basin and Range. The recent research of Hill and Holliday (2011) expands on earlier research conducted in the Great Basin focusing on early forager-landscape interactions in paleolake basins (Beck and Jones 1990, 1997; Jones et al. 2003). These studies, in addition to studies conducted in the Rio Grande (Judge 1973) and Tularosa

(Wessel et al. 1997) Basins, point toward a Paleoindian subsistence focus on littoral resources—in contrast to a less specific paleoenvironmental focus for later occupations. However, inferences about spatio-temporal trends in land use informed by the distribution of surface sites remain speculative in the absence of a concrete temporal framework; each cultural component may span centuries to millennia, and individual components (Clovis, Folsom, and early Archaic) may actually be comprised of one or more short-term occupations associated with periods of atypical paleoenvironmental conditions.

Langford (2003) found sites along and above higher shorelines in the Tularosa and Otero Basins; however, shoreline ages remain unknown and possible preservation biases must be taken into account, particularly since basin floors were subject to deflation in the early Holocene. Weber (1994) suggested that Clovis artifacts in the San Agustin Basin remain buried within playa and marsh muds along the highest shorelines, and that Folsom artifacts remain buried within eolian deposits associated with lower shorelines in the White Lake sub-basin. Holliday et al. (2006) and Weber (1980) suggest that Folsom age deposits are likely preserved at the Ake site, near a lower shoreline in the C–N sub-basin. Hill and Holliday (2011) demonstrated that there is no obvious correlation between archaeological sites and paleoshorelines in the C–N and White Lake sub-basins, except for a series of Archaic and Folsom components associated with a shoreline ridge at 2,105 m in the former. Based on the density and distribution of surface components, they suggested that environmental conditions in the C–N were always more favorable for habitation—and thus conducive to longer occupations—than in the White Lake. Accordingly, they suggested that the lack of Clovis site clustering along paleoshorelines in either sub-basin reflects sparse or no inhabitation during initial colonization of the region. They tentatively concluded that Folsom foragers were camping along the 2,105 m paleoshoreline in the C–N sub-basin to exploit resources that were not previously equal in abundance. Finally, as conditions became drier and permanent lakes eventually disappeared, late Paleoindian and early Archaic foragers exploited an increasingly broad array of paleoenvironments (Hill and Holliday 2011).

DiPeso (1953) reported a single Clovis point from a large dune adjacent to the northeastern margin of Willcox Playa, found on the erosional surface of a recently exhumed paleosol. Extensive archaeological survey data from Willcox Basin reveals that Archaic surface sites predominantly occur in playa margin and higher upland settings (Waters and Woosley 1990), but a lack of information about Paleoindian occupation of the area precludes us from conducting a more detailed analysis following Hill and Holliday (2011). In addition, the lake probably never again topped the beach ridge following its construction between 17 and 14 Ka. Therefore, Paleoindian or early Archaic foragers might have encountered lowstand lakes whose shoreline features were subsequently obliterated by wind deflation. Nonetheless, it is clear from the stratigraphic record of the San Pedro Valley that P:E increased abruptly at ~12.8 Ka, and from that we might expect to find evidence for synchronous changes in the nearby Willcox Basin.

1.7.2 Regional

Considerable debate has surrounded the causes of megafaunal extinctions, contemporaneous with the brief appearance of Clovis in the stratigraphic record of southeastern AZ (Ballenger et al. 2011). Several explanations have been advanced, most notably overhunting (Martin 1967) and an abrupt freeze following an "extraterrestrial" (cometary) impact (Firestone et al. 2007). Based on stratigraphic evidence, Haynes (1991) advanced the "Clovis Drought" hypothesis, contending that humans arrived in the region during the peak of Bølling-Allerød aridity, immediately preceding the abrupt onset of cooler/wetter conditions at the beginning of the Younger Dryas chronozone, as informed by ^{14}C dating of the Black Mat (Haynes 2007). Haynes (1991) posited that by this time, diminishing spring activity caused animals to congregate around shrinking watering holes, where people could have hunted them with ease. However, it is unclear whether the extinction occurred before, or after the abrupt shift to wetter conditions throughout the region shortly after 13 Ka—or whether or not the Younger Dryas was immediately preceded by a brief period of extreme aridity. A more complete paleohydrologic record from fossil wetland and shoreline deposits within the region could provide this information.

1.8 Summary and Conclusions

Collectively, fossil shorelines represent a largely untapped source of paleohydrologic information that would complement proxy records of lake-level change derived from sediment cores, in addition to groundwater records from spring-discharge deposits. Because they provide absolute constraints on the timing and extent of past lake expansions, they are critical for improving current knowledge about the hydrologic and geomorphic evolution of ancient landscapes in response to paleoclimatic variability. In turn, knowledge about the timing of lake-level fluctuations enables archaeologists to assess how intensively ancient occupants of the region might have utilized lake- and playa-margin resources during the Pleistocene-Holocene transition, as the supply of surface water and ecological resources dwindled. Finally, paleoshorelines can provide geomorphic context for interpreting the significance of spatial patterns in the archaeological record of individual basins.

Ballenger et al. (2011) drew attention to correlations between paleoclimate changes exhibited by global and regional records, and major shifts in human subsistence strategies inferred from the archaeological record of the southeastern Basin and Range. They suggested that the following can be explained by broad-scale paleoenvironmental and paleoecological changes: (1) the brief appearance of Clovis in the stratigraphic record after 13 Ka, (2) subsequent extinction of mammoth and a subsistence shift to exploitation of bison herds in adjacent parts of the Plains by 12.7 Ka, (3) a hiatus in the archaeological record from 12.7–11.7 Ka, (4) increased occupational intensity of the Plains by Folsom inhabitants beginning at 12.7 Ka, and (5) the appearance of Archaic adaptations into southeastern AZ after 11.7 Ka.

Records of lake-level change in the region are currently incomplete or unreliable, and therefore cannot inform us about the timing and expression of paleohydrologic changes likely accounting for major changes in Paleoindian subsistence and mobility during the terminal Pleistocene and early Holocene (Ballenger et al. 2011). At present, ^{14}C dating of shoreline deposits shows that short-lived lake stands occurred between 15 and 13 Ka, warranting modification of the drought scenario proposed by Polyak et al. (2012). Paleohydrologic nuances of this long-lived drought are provided by the San Pedro and Willcox Basin paleohydrologic records, the former of which indicates an unprecedented drop in water tables by ~15 Ka (Long 1966; Haynes 2007; Pigati 2009). Stratigraphic evidence from Murray Springs in the San Pedro Basin indicates that vigorous and probably intermittent spring discharge occurred throughout the Bølling-Allerød, well into Clovis times—when springs apparently ran dry for a very brief period before an abrupt rise in water tables throughout western North America at ~12.8 Ka (Haynes 2008). If discovered, a simultaneous expansion of lakes in the southeastern Basin and Range would allow for quantification of the P:E increase that caused this hydrologic rebound. In conclusion, continued investigation of fossil shoreline deposits in southern AZ and NM is a necessary step toward reaching definitive conclusions about the climatic factors effecting dramatic changes in hydrology, ecology, and human subsistence and mobility within the southeastern Basin and Range during the terminal Pleistocene-early Holocene transition.

References

Adams DK, Comrie AC (1997) The North American monsoon. Bull Am Meteorol Soc 78(10):2197–2213

Allen BD (2005) Ice age lakes in New Mexico. In: Lucas SG, Morgan GA, Zeigler KE (eds) New Mexico's ice ages. New Mex Mus Nat Hist Sci Bull 18:107–114

Allen BD, Anderson RY (2000) A continuous, high-resolution record of late Pleistocene climate variability from the Estancia Basin, New Mexico. Geol Soc Am Bull 112(9):1444–1458

Alley RB (2000) The Younger Dryas cold interval as viewed from central Greenland. Quatern Sci Rev 19:213–226

Anderson RY, Allen BD, Menking KM (2002) Geomorphic expression of abrupt climate change in southwestern North America at the Glacial termination. Quatern Res 57:371–381

Antevs E (1954) Climate of New Mexico during the last glacio-pluvial. J Geol 62:182–191

Asmerom Y, Polyak VJ, Burns SJ (2010) Variable winter moisture in the southwestern United States linked to rapid glacial climate shifts. Nat Geosci 3:114–117

Ballenger JAM (2010) Late Quaternary paleoenvironments and archaeology in the San Pedro Basin, southeastern Arizona, U.S.A. Unpublished Ph.D. Dissertation, Department of Anthropology, University of Arizona, Tucson, 205 pp

Ballenger JAM, Holliday VT, Kowler AL, Reitze WT, Prasciunas MM, Miller DS, Windingstad JD (2011) Evidence for Younger Dryas global climate oscillation and human response in the American Southwest. Quatern Int 242:502–519

Beck C, Jones GT (1990) Toolstone selection and lithic technology in early Great Basin prehistory. J Field Archaeol 17:283–299

Beck C, Jones GT (1997) The terminal Pleistocene/early Holocene archaeology of the Great Basin. J World Prehist 11:161–236

Benson LV, Paillet FL (1989) The use of total lake-surface area as an indicator of climatic change: examples from the Lahontan Basin. Quatern Res 32:262–275

Benson LV, Currey DR, Dorn RI, Lajoie KR, Oviatt CG, Robinson SW, Smith GI, Stine S (1990) Chronology of expansion and contraction of four Great Basin lake systems during the past 35,000 years. Palaeogeogr Palaeoclimatol Palaeoecol 78:241–286

Brackenridge GR (1978) Evidence for a cold dry Full-Glacial climate in the American Southwest. Quatern Res 9:22–40

Broecker WS, Walton AS (1959) The geochemistry of ^{14}C in fresh water systems. Geochim Cosmochim Acta 16:15–38

Connell SD, Hawley JW, Love DW (2005) Late Cenozoic drainage development in the southeastern Basin and Range of New Mexico, southeasternmost Arizona, and western Texas. In: Lucas SG, Morgan GS, Zeigler KE (eds) New Mexico's ice ages. New Mex Mus Nat Hist Sci Bull 28:125–150

Deevey ES, Gross MS, Hutchinson GE, Kraybill HI (1954) The natural ^{14}C content of materials from hard-water lakes. Proc Natl Acad Sci 40:285–288

DiPeso CC (1953) Clovis fluted points from southwestern Arizona. Am Antiq 19:82–85

Firestone RB, West A, Kennett JP, Becker L, Bunch TE, Revay ZS, Schultz PH, Belgya T, Kennett D, Erlandson JM, Dickenson J, Goodyear AC, Harris RS, Howard GA, Kloosterman JB, Lechler P, Mayewski PA, Montgomery J, Poreda R, Darrah T, Que Hee SS, Smith AR, Stich A, Topping W, Wittke JH, Wolbach WS (2007) Evidence for an extraterrestrial impact 12,900 years ago that contributed to the megafaunal extinctions and the Younger Dryas cooling. Proc Natl Acad Sci 104:16016–16021

Fleischauer HL Jr, Stone WJ (1982) Quaternary geology of Lake Animas, Hidalgo County, New Mexico. New Mex Bur Mines Min Resour Circular 174:25

Galloway RW (1983) Full-glacial southwestern United States: mild and wet or cold and dry? Quatern Res 19:236–248

Hall SA, Penner WL, Palacios-Fest MR, Metcalf AL, Smith SJ (2012) Cool, wet conditions late in the Younger Dryas in semi-arid New Mexico. Quatern Res 77:87–95

Hawley JW (1993) Geomorphic setting and late Quaternary history of pluvial lake basins in the southern New Mexico region. New Mex Bur Mines Min Resour Open-File Rep 391:28

Haynes CV Jr (1991) Geoachaeological and paleohydrological evidence for a Clovis-age drought in North America and its bearing on extinction. Quatern Res 35:438–450

Haynes CV Jr (2007) Quaternary geology of the Murray Springs Clovis site. In: Haynes CV Jr, Huckell BB (eds) Murray Springs: a Clovis site with multiple activity areas in the San Pedro Valley, Arizona. Anthropology papers of the University of Arizona, vol 71. University of Arizona Press, Tucson, pp 16–54

Haynes CV Jr (2008) Younger Dryas "Black Mats" and the Rancholabrean Termination in North America. Proc Natl Acad Sci 105(18):6520–6525

Haynes CV Jr, Long A, Jull AJT (1987) Radiocarbon dates at Willcox Playa, Arizona, bracket the Clovis occupation surface. Curr Res Pleistocene 4:124–126

Hill ME Jr, Holliday VT (2011) Paleoindian and later occupations along ancient shorelines of the San Agustin Plains, New Mexico. J Field Archaeol 36(1):3–20

Holliday VT, Weber RH, Mayer JH (2006) Geoarchaeological investigations at the Mockingbird Gap Folsom site, New Mexico. Curr Res Pleistocene 23:112–114

Jones GT, Beck C, Jones EE, Hughes RE (2003) Lithic source use and Paleoarchaic foraging territories in the Great Basin. Am Antiq 68:5–34

Judge WJ (1973) Paleoindian occupation of the central Rio Grande Valley in New Mexico. University of New Mexico Press, Albuquerque

Krider PR (1998) Paleoclimatic significance of late Quaternary lacustrine and alluvial stratigraphy, Animas valley, New Mexico. Quatern Res 50:283–289

Langbein WB (1961) Salinity and hydrology of closed lakes. U.S. Geological Survey Professional Paper 412. Government Printing Office, Washington, 20 pp.

Langford RP (2003) The Holocene history of the White Sands dune field and influences on eolian deflation and playa lakes. Quatern Int 104:31–39

Leopold LB (1951) Pleistocene climate in New Mexico. Am J Sci 249:152–168

Long (1966) Late Pleistocene and recent chronologies of playa lakes in Arizona and New Mexico. Unpublished Ph.D. Dissertation, Department of Geosciences, University of Arizona, Tucson, 141 pp

Markgraf V, Bradbury JP, Forester RM, McCoy W, Singh G, Sternberg RS (1983). Paleoenvironmental reassessment of the 1.6 million-year-old record from San Agustin Basin, New Mexico. In: Chapin EC, Callender JF (eds) New Mexico Geological Society 34th annual field conference guidebook, pp 291–297

Markgraf V, Bradbury JP, Forester RM, Singh G, Sternberg RS (1984) San Agustin Plains, New Mexico: age and paleoenvironmental potential reassessed. Quatern Res 22:336–343

Martin P (1967) Prehistoric overkill. In Pleistocene extinctions: the search for a cause. Yale University Press, New Haven, pp 75–120

Meinzer OE (1922) Map of the Pleistocene lakes of the Basin and Range province and its significance. Geol Soc Am Bull 33:541–552

Meinzer OE, Kelton FC (1913) Geology and water resources of Sulphur Spring Valley, Arizona. U.S. Geological Survey Water-Supply Paper 320. Government Printing Office, Washington, p 231

Mifflin MD, Wheat MM (1979) Pluvial lakes and estimated pluvial climates of Nevada. Nevada Bur Mines Geol Bull 94:57

Phillips FM, Campbell AR, Kruger C, Johnson PS, Roberts R, Keyes E (1992) A reconstruction of the response of the water balance in Western United States lake basins to climate change. New Mexico Water Resources Research Institute, Technical Completion Report (vol 1) on Project Nos. 11345662 and 1423687, WRRI Report No. 269, 167 pp

Pigati JS, Bright JE, Shanahan TM, Mahan SA (2009) Late Pleistocene paleohydrology near the boundary of the Sonoran and Chihuahuan Deserts, southeastern Arizona, USA. Quat Sci Rev 28:286–300

Pigati JS, Quade J, Shanahan TM, Haynes CV Jr (2004) Radiocarbon dating of minute gastropods and new constraints on the timing of late Quaternary spring-discharge deposits in southern Arizona, USA. Palaeogeogr Palaeoclimatol Palaeoecol 204:33–45

Pigati JS, Rech JA, Nekola JC (2010) Radiocarbon dating of small terrestrial gastropod shells in North America. Quat Geochronol 5:519–532

Polyak VJ, Asmerom Y, Burns SJ, Lachniet MS (2012) Climatic backdrop to the terminal Pleistocene extinction of North American mammals. Geology 40(11):1023–1026

Polyak VJ, Rasmussen JBT, Asmerom Y (2004) Prolonged wet period in the southwestern United States through the Younger Dryas. Geology 32:5–8

Powers WE (1939) Basin and shore features of extinct Lake San Agustin, New Mexico. J Geomorphol 2:345–356

Rech JA, Pigati JS, Lehmann CN, McGimpsey DA, Grimley DA, Nekola JC (2011) Assessing open-system behavior of carbon-14 in terrestrial gastropod shells. Radiocarbon 53(2):225–325

Reimer P, Baillie L, Bard E, Bayliss A, Beck W, Blackwell G, Bronk Ramsey C, Buck E, Burr S, Edwards L, Friedrich M, Grootes M, Guilderson P, Hajdas I, Heaton J, Hogg G, Hughen A, Kaiser F, Kromer B, McCormac G, Manning W, Reimer W, Richards A, Southon R, Talamo S, Turney M, van der Plicht J, Weyhenmeyer E (2009) IntCal 09 and Marine 09 radiocarbon age calibration curves, 0–50,000 years cal BP. Radiocarbon 51:1111–1150

Robinson RC (1965) Sedimentology of beach ridge and nearshore deposits, pluvial Lake Cochise, southeastern Arizona. Unpublished M.S. Thesis, Department of Geosciences, University of Arizona, Tucson, 111 pp

Schreiber, JF (1978) Geology of the Willcox Playa, Cochise County, Arizona. In: Callender JF, Wilt JC, Clemons RE (eds) Land of Cochise. New Mexico Geological Society 29th annual field conference guidebook, pp 277–282

Schwennesen AT (1918) Ground water in the Animas, Playas, Hachita, and San Luis Basins, New Mexico. U.S. Geological Survey Water-Supply Paper 422. Government Printing Office, Washington, 35 pp

Smith LN, Anderson RY (1982) Pleistocene-Holocene climate of the Estancia Basin, central New Mexico. In: Grambling JA, Wells SG, Callender, JF (eds) Albuquerque country II. New Mexico Geological Society 33rd annual field conference guidebook, pp 347–350

Stuiver M, Reimer RW (1993) Extended ^{14}C database and revised CALIB radiocarbon calibration program. Radiocarbon 35:215–230

Wagner JDM, Cole JE, Beck JW, Patchett PJ, Henderson GM, Barnett HR (2010) Moisture variability in the southwestern United States linked to abrupt glacial climate change. Nat Geosci 3:110–113

Waters MR (1989) Late Quaternary lacustrine history and paleoclimate significance of pluvial Lake Cochise, southeastern Arizona. Quatern Res 32:1–11

Waters MR, Woosley AI (1990) The geoarchaeology and preceramic prehistory of the Willcox Basin, SE Arizona. J Field Archaeol 17(2):163–175

Weber RH (1980) Geology of the Ake Site. In: Beckett PH (ed) The Ake Site: collection and excavation of LA13423, Catron County, New Mexico. New Mexico State University cultural resources management division report 357. Department of Sociology and Anthropology, New Mexico State University, Las Cruces, pp 221–238

Weber RH (1994) Pluvial lakes of the Plains of San Agustin. In: Chamberlin RM, Kues BS, Cather SM, Barker JM, McIntosh WC (eds) Mogollon Slope, West-Central New Mexico and east-central Arizona. New Mexico Geological Society 45th annual field conference guidebook, pp 9–11

Wessel RL, Eidenbach PL, Meyer LM, Comer CS, Knight B (1997) From playas to highlands: Paleo-Indian adaptations to the region of the Tularosa. Human Systems Research, Inc., Tularosa

Chapter 2
Post-Mazama River Terraces and Human Occupation Along the North Umpqua River Oregon

Dorothy E. Freidel and Brian L. O'Neill

Abstract A sequence of three post-Mazama terraces, found at intervals along the banks of the North Umpqua River, western Oregon, contains evidence of middle and late Holocene human occupation that has contributed to a construction of a geomorphic history of terrace formation. The evidence suggests that different geomorphic mechanisms were responsible for the development of the three terraces. It also verifies that people were living along the North Umpqua River before the climactic eruption of Mount Mazama and may have returned within a few hundred years after. The highest, Panther Terrace, formed from the glowing ash flow deposits from the Mount Mazama eruption, ca 7600 years B.P. The next lower Illahee Terrace developed through slow aggradation of the gravelly channel bed followed by an episode of rapid deposition of reworked Mazama ash by around 4500 cal. yrs B.P. The lowest Eagle Terrace formed on a possible pre-Mazama strath terrace within the last 2000 years, and was probably abandoned within the past few hundred years. Pollen records from the Cascades suggest a possible climatic stimulus for the development and abandonment of the lower two terraces.

Keywords River terraces • Mazama eruption • Geoarchaeology • North Umpqua River • Oregon

2.1 Introduction

This research investigates the origin and mechanisms of formation of three post-Mazama terraces found at intervals along the rock-bound gorge of the North Umpqua River in the western Cascades, Oregon. A major hypothesized cause of terrace

D. E. Freidel (✉)
Department of Geography, Sonoma State University, Rohnert Park, CA 94928, USA
e-mail: freidel@sonoma.edu

B. L. O'Neill
Oregon State Museum of Anthropology, University of Oregon, Eugene, OR 97403, USA

M. Goman, *Human Environment Interactions – Volume 2*,
SpringerBriefs in Earth System Sciences, DOI: 10.1007/978-3-642-36880-6_2,

construction was the response of the river to the abrupt input of large quantities of pyroclastic debris that poured into the river channel during the climactic eruption of Mount Mazama, 7600 years B.P. [All radiometric ages referred to in this paper are calibrated (Stuiver and Reimer 2011).] Radiometric ages and stratigraphic context of archaeological materials, along with sedimentological characteristics and soil development in terrace deposits, help define the history and geomorphic mechanisms that created these terraces. Archaeological sites, and the presence versus absence of cultural features, may provide evidence of variations in human occupation of the region before and after the eruption, or may reflect times when river flooding made its banks uninviting to people. For ease of discussion, the three terraces, from highest to lowest, are referred to as the Panther, Illahee, and Eagle terraces (O'Neill et al. 1996).

To briefly clarify the geomorphic terms, when a river channel is aggrading, it deposits sediment within and adjacent to the channel during high discharge events. These deposits become fill terraces when the river begins to incise, or cut down into the bed of its channel. At this point, even during flooding, the river discharge no longer reaches high enough to deposit sediments on the old floodplain, and the terrace surface is abandoned. It thus becomes a stable surface available for habitation. When abandonment of the old floodplain occurs a new lower active channel level is established. A strath surface is formed when the river channel migrates laterally across the river plain, eroding the sediments and rock surface beneath the channel. A strath terrace or bench is formed when the channel begins to incise, eroding its bed, rather than laterally migrating. Thus fill terraces are depositional and strath surfaces are erosional.

Geomorphological and archaeological studies are useful partners in the investigation of river terraces. River terraces are valuable in archaeological investigations because they represent stable surfaces that were available for human occupancy during discrete time periods (cf. Holliday 1987; Gardner and Donahue 1985; Bettis and Benn 1984; Brakenridge 1984). Archaeological sites commonly provide datable material that may otherwise be rare in sediments underlying terrace surfaces, contributing to understanding the geomorphic chronology of terrace development. On the other hand, geomorphically correlated and well-dated terrace surfaces that can be traced for some distance downstream can provide maximum ages on cultural sites and assist in temporal correlation between sites. The question of what paleoenvironmental changes the terraces signify is also of importance in understanding the circumstances of the ancient inhabitants of this riverine setting.

Alluvial terraces have been studied as potential indicators of periods of instability and environmental change (Bull 1991; Schumm and Brakenridge 1987). Rivers adjust their channels and floodplains by cutting and filling in response to changes in sediment load, discharge, or stream gradient. The major stimuli for these periods of river adjustment are thought to be climate change and/or tectonic movement. Both are likely mechanisms at this location, with ongoing uplift of the Cascade Range at least partially responsible for terrace development over time. Uplift in this section of the Cascades has been estimated at about 0.2 ± 0.3 mm/yr, which would indicate about 15.2 cm uplift since the Mazama eruption (Personius 1995). Once equilibrium is disrupted, further cut and fill sequences can occur without additional external forcing.

Thus terraces may continue to form as a result of complex response to intrinsic thresholds within the river system (Schumm 1977, 1979, 2005). Volcanic eruptions, with sudden inputs of ash, pumice, and coarser ejecta into drainages, also disrupt river equilibrium by dramatically increasing the sediment load. This latter stimulus occurred along the North Umpqua River during the climactic eruption of Mount Mazama, 7627 ± 150 calibrated years B.P. (Zdanowicz et al. 1999, Adams 1990). [Unless otherwise indicated, ages provided herein are dendrocalibrated (Stuiver and Reimer 2011). Obsidian hydration ages are calculated using dendrocalibrated radiocarbon ages.] It should be noted that it is often difficult to interpret terraces as evidence of a particular geomorphic history because of discontinuous terrace remnants along a channel, the difficulties with correlation from one terrace remnant to the next, and the range of possible stimuli for disequilibrium, whether external events or intrinsic threshold controls (Ritter et al. 2011, pp. 280–282). In this study we relied chiefly on the unique characteristics of sediment units found within each terrace as well as their relative elevation above the modern channel.

Initially, we hypothesized that the complex response of the river to the abrupt change in sediment load resulting from the eruption had caused the development and subsequent abandonment of the three terraces. Yet, examination of the morphology of the terraces and the timing of their formation and abandonment suggests that other factors, such as climate variation, may have also been influential. If climate variation can be documented as a cause of regional fluvial adjustments, this could have significance for other archaeological sequences in western Oregon.

The major questions we have considered in this study are the following:

1. What environmental mechanisms stimulated the formation of the three terraces? Was terrace formation primarily a response to the ash flows from the Mazama eruption? Were other mechanisms involved, such as complex response (feedbacks within the river system), climate variation, or tectonic adjustment (uplift, subsidence)?
2. Does the temporal distribution of cultural features on the terraces reflect timing of terrace development (availability of stable land surfaces)?

2.2 The Study Area

The North Umpqua River and its tributaries (principally the Clearwater River, Steamboat Creek, Fish Creek, and Boulder Creek) drain 3502 km^2 of the western Cascades east of Roseburg, Oregon (Fig. 2.1). The headwaters begin at an elevation of about 1800 m in the relatively gentle uplands of Tertiary and Quaternary lava flows and pumice plains of the High Cascades Physiographic province north of Crater Lake (Franklin and Dyrness 1973; Baldwin 1981). These uplands were scoured by glaciers in the middle and late Pleistocene, leaving behind morainal debris and glacial outwash. Thick airfall tephra deposits laid down during the Plinean eruption of Mount Mazama blanket large areas of the old lava flows and

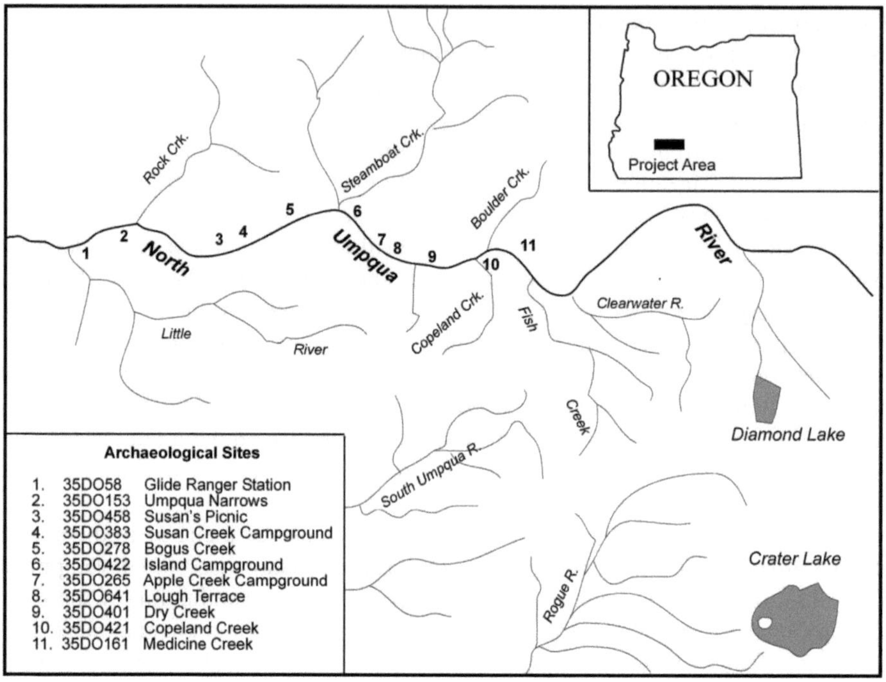

Fig. 2.1 Topographic map of study area, including headwaters area of North Umpqua, and downstream as far as Susan Creek. All sites referred to in the text are indicated on the map, including Medicine Creek, Apple Creek Campground, Bogus Creek, Susan Creek, and Glide

glacial deposits (Fig. 2.2, MacLeod et al. 1995; Bacon 1983). Descending the western slopes of the Cascades, the North Umpqua and other rivers have deeply incised the older volcanics—to the south, Pliocene and Pleistocene basalt and andesitic-covered uplands, and to the north, Miocene volcanics and areas of mid-Pleistocene basaltic andesite flows (Sherrod 1991).

Flowing from its headwaters near the crest of the southern Cascades, the North Umpqua River has cut a steep canyon over 250 m deep in places, with a number of cascades and waterfalls. These mountainous reaches of the modern North Umpqua are for the most part bedrock-controlled and bedload-dominated. The channel averages several tens of meters across with occasional gravel bars, rapids, and deep bedrock pools. Fluvial terraces have formed intermittently along the channel, commonly within meander bends and downstream of bedrock constrictions in the channel where the gorge widens.

At the upper end of the study area, the channel falls 132 m in 21 km, from an elevation of 482 m near the confluence of Copeland Creek to 350 m at Island Campground, with a mean gradient of 0.006. Between Island Campground and Susan Creek Campground, 18 km downstream at 274 m, the gradient flattens to 0.004. Mean discharge measured at a gaging station just upstream from Copeland Creek is 41.3 m^3ps, averaged over the period of record from

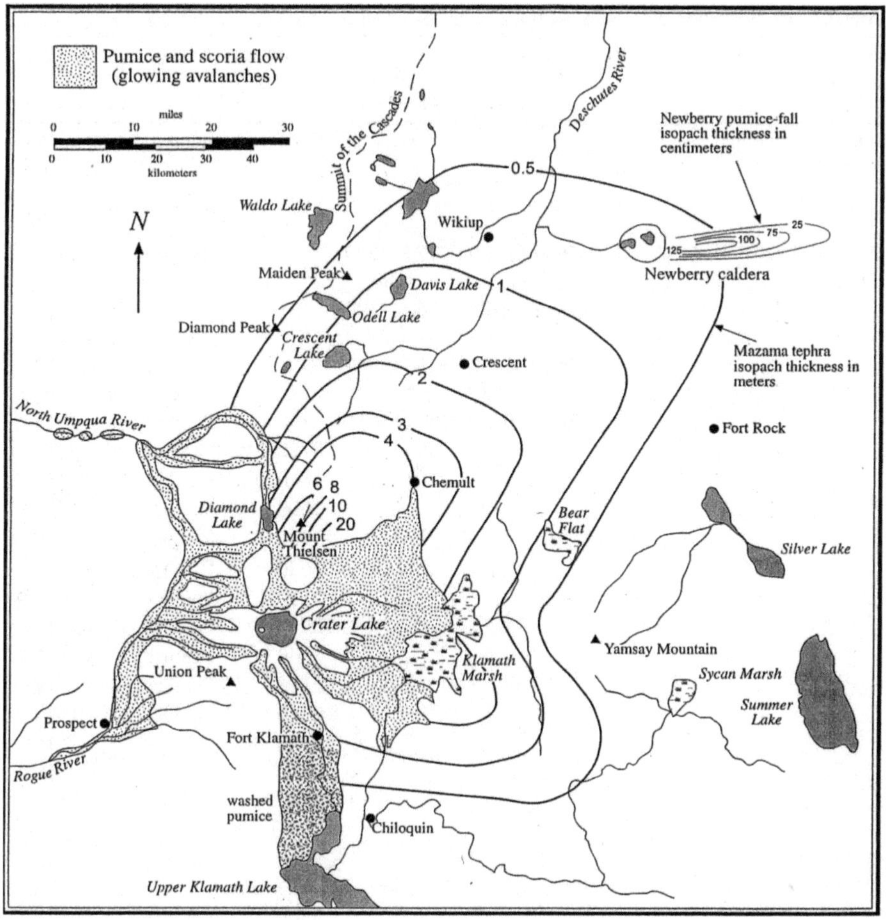

Fig. 2.2 Map showing distribution of Mazama airfall tephra and pyroclastic flows. Note on the upper left the ash flow deposits along the North Umpqua River (MacLeod et al. 1995)

1950–2011 (U.S. Geological Survey 2012). Maximum discharge for the period of record was estimated at 1152 m^3ps on December 22, 1964, and minimum discharge was measured at 9.3 m^3ps on July 11, 2004 (U.S. Geological Survey 2011).

The climate in the study area is characterized by mild, wet winters, with temperatures averaging 4 °C. Summers are warm and generally dry, with mean temperatures of 20 °C and occasional frontal or convective precipitation. Annual precipitation varies from about 1015–1525 mm, primarily as winter and spring rain at the lower elevations. Above 300 m elevation, light snow may persist for several weeks during the winter, and significant snowfalls are received at higher elevations in the vicinity of the study area (Wert et al. 1977, p. 25).

The archaeological sites investigated in the present study include, from upstream to downstream, Copeland Creek (35DO421), Dry Creek (35DO401),

Table 2.1 Radiocarbon dates and calibrated ages by stratigraphic unit and terrace

Site number/Name	14 C (BP)	Lab No.	Calibrated age/range (BP)	Hydration Age (BP)	Terrace	Unit
35DO421/Copeland Creek				2990	Illahee	IV
				625	Illahee	IV
35DO401/Dry Creek	6630 ± 180	Beta-50250	7770 (7500) 7180		Panther	I
	6540 ± 170	Beta-48725	7650 (7390) 7040		Panther	I
	5220 ± 80	Beta-31832	6190 (5980) 5760		Panther	I-IV
	130 ± 50	Beta-50250	290 (130) 0		Panther	IV
				8780	Panther	I
				7840	Panther	I
				7600	Panther	I
				5860	Panther	IV
				5620	Panther	IV
				4510	Panther	IV
				3710	Panther	IV
				3340	Panther	IV
				2500	Panther	IV
				2340	Panther	IV
				1520	Panther	IV
				980	Panther	IV
				550	Panther	IV
35DO641/Lough Terrace				8390	Panther	I
				4300	Panther	IV
				1400	Panther	IV
35DO265/Apple Creek Camp	3500 ± 100	Beta-34382	4080 (3790) 3470		Illahee	IV
				6300	Illahee	III?
				5800	Illahee	III?
				3500	Illahee	IV
35DO422/Island Camp	4050 ± 60	Beta-48726	4815 (4460) 4405		Illahee	IV
	1880 ± 110	Beta-50253	2055 (1820) 1540		Illahee	IV
	1210 ± 70	Beta-34384	1285 (1110) 960		Panther	II-IV
				1910	Panther/Illahee	II-IV
				880	Panther/Illahee	II-IV

(continued)

Site number/Name	14 C (BP)	Lab No.	Calibrated age/range (BP)	Hydration Age (BP)	Terrace	Unit
35DO383/Susan Creek	6890 ± 90[a]	Beta-49629	7900 (7600) 7540		Illahee	III
	6840 ± 70[a]	Beta-58852	7760 (7630) 7530		Illahee	III
	6790 ± 60[a]	Beta-58849	7680 (7580) 7480		Illahee	III
	4030 ± 60	Beta-58850	4810 (4470) 4350		Illahee	IV
	1190 ± 80	Beta-58847	1280 (1070) 940		Illahee	IV
	1120 ± 50	Beta-58846	1160 (1040) 930		Illahee	IV
	1010 ± 70	Beta-58848	1060 (930) 750		Illahee	IV
	1000 ± 60	Beta-58851	1050 (930) 770		Illahee	IV

[a]Dates on non-cultural charcoal, from trees incinerated during the Mt. Mazama eruption, in ash flows

Lough Terrace (35DO641), Island Campground (35DO422), and Susan Creek Campground (35DO383) (Fig. 2.1). With the exception of the Copeland Creek site, all of the sites described in this study are on the north bank of the North Umpqua River. This is not because of a lack of cultural occupation on the south bank, nor a lack of matching (paired) terraces, but primarily because of the need for survey and mitigation for highway construction. We incorporate some data from the Medicine Creek (35DO161), Narrows (35DO153), Apple Creek Campground (35DO265), and Glide Ranger Station (35DO58) archaeological sites.

2.3 Methods

The topography and geology of the study area were examined on USGS 7.5' topographic quads, the geologic map of the central Cascades (Sherrod 1991), and air photos taken by the U.S. Forest Service along the North Umpqua River channel. At each site, sediments and soil development were described in archaeological test pits, backhoe trenches, and auger holes using NRCS soil criteria and terminology (Schoeneberger et al. 1998). Sediment samples were collected for lab examination, including particle size analysis of selected sample sets (Gee 2002). The archaeological teams mapped their excavations on the terraces with transit and rod, usually to an arbitrary 100-m base elevation, and determined the elevation of the terraces by surveying to the water's edge. Combined GPS and laser range finder survey of terrace elevations at three of the sites (Dry Creek, Island, and Susan's Creek) was conducted.

Table 2.2 Description of stratigraphic units

Unit	Age (Cal yr BP)	Composition	Terrace	Thickness
I	>7600	Pre-Mazama paleosol in gravelly colluvium, dark reddish brown (7.5YR) sandy clay loam; moderate soil development, 70 cm thick Bt horizon	Panther, Illahee Eagle	Varies
II	7600	Primary Mazama tephra ash flow, ash cloud fine, massive gray (2.5Y-10YR), loamy sand with pumice lapilli, blocks, weak soil development diagnostic subangular pumice lapilli layer 2-5 cm at base of primary ash	Panther,	0.18- > 15 m
III	>7600- ~4500	Bedded sand & gravels, channel deposits, glacial outwash & colluvium interbedded with reworked Mazama pumice and ash; little soil development observed	Illahee	1-4 m
IV	4500- ~3800	Reworked Mazama tephra overbank flood deposits; yellowish brown (10YR 5/6) loamy sand with pumice lapilli, blocks; weak soil development	Illahee (Panther?)	0.6- > 5.5 m
V	~1000	Recent floodplain sediments fine dark brown sandy clay loam, (10 YR 3/4), no tephra, weak soil development	Eagle	<1 m
VI	<7600	Post-Mazama colluvium dark reddish brown (7.5YR 3/4) clay loam	Panther	<1 m

The individual stratigraphic units were defined and correlated between sites on the basis of sediment characteristics, soil development, stratigraphic relationships, geomorphic context, and radiometric dating (Tables 2.1, 2.2, 2.3). Age control for the development of the terrace deposits was obtained from radiocarbon assay on charcoal associated with cultural features, obsidian hydration measurements on cultural artifacts, and the presence or absence of primary Mazama tephra (Table 2.1).

Obsidian tools and waste flakes of two central Oregon obsidian sources are commonly found in Umpqua Basin archaeological deposits: Silver Lake/Sycan Marsh and Spodue Mountain (Musil and O'Neill 1997). Both have been found to hydrate at approximately the same rate, calculated at 4.1 μ^2/1000 years (O'Neill 2002; Connolly 1991). Archaeologically collected specimens, those with known

Table 2.3 Elevations and ages of North Umpqua Terraces

Terrace	Elev. above modern channel (m)	Stratigraphic Units	Period(s) of aggradation cal yrs BP	Timing of abandonment cal yrs BP
Panther	15–25 m[a]	I, II, IV, VI	7600	<7600
Illahee	10.5–14 m[b]	III, IV	<7600–4500	4500–3800
Eagle	5–8 m[c]	V	3000–1100	<1100 possible historic flood

[a]25 m at Copeland Creek, elsewhere consistently ~15–16 m
[b]Increases from 10.5 m at Copeland Creek to 14 m at Island, then consistent at 14 m downstream
[c]5 m at Island, otherwise consistently ~7–8 m both up and downstream

provenience and stratigraphic context, were submitted for sourcing and hydration studies (O'Neill 1991, 2002; O'Neill et al. 1996). The results of these analyses provided calculated hydration ages that were used to help determine limiting ages for the sediments within which the obsidian specimens were discovered.

Correlation of the tephra deposits with the climactic eruption of Mount Mazama was established by Foit (1991), Geoanalytical Lab, Washington State University, using electron microprobe analysis. One tephra sample collected during data recovery investigations at the Dry Creek archaeological site was submitted for geochemical analysis to confirm its origin. The sample was collected from an approximately 2–5 cm thick layer of angular gravel-sized pumice lapilli found at the interface between a paleosol and a fine gray ash at approximately 95 cm below the surface. The analysis found the Dry Creek sample to closely match volcanic glass from the climactic eruption of Mount Mazama, establishing the identity of the tephra's origin and confirming the antiquity of the cultural remains buried in the underlying paleosol (O'Neill et al. 1996). This analysis is supported by numerous radiocarbon assays on charcoal associated with the volcanic ash in the study area that correspond with the Mazama eruption (Table 2.1).

2.4 Terrace Formation

The eruption of Mount Mazama has had a profound influence on the Holocene geomorphic history of the North Umpqua River. Shortly following the Plinean eruption of Mount Mazama, glowing avalanches of pyroclastic debris flowed down the sides of the volcano in all directions (Fig. 2.2). Part of this ash flow entered the North Umpqua River drainage at its headwaters near Diamond Lake. The ash clogged the river channel far downstream, as much as 25 m deep at Copeland Creek, over 50 river km from the volcano; more than 18 m deep at Dry Creek, 57 km downstream; 20 m deep at Lough Terrace, 64 km downstream; and 16 m deep at Island Campground, 75 km downstream (Fig. 2.1). Today, virtually all of this sediment has been removed or stabilized, much of it stored as primary or reworked pyroclastic deposits constructing the three river terraces described herein (Table 2.2). More than ten archaeological sites, with cultural components ranging from early Holocene pre-Mazama sites to middle and late Holocene campsites, have been found beneath, within, and on the terrace surfaces in the study area. Datable materials recovered from these components have helped provide age control on the timing of terrace formation.

2.4.1 Stratigraphic Units

The terraces are underlain by six stratigraphic units (Table 2.2). (See full text of sediment and soil descriptions in O'Neill et al. 1996.)

Unit I. Oldest is Unit I, a dark reddish brown (7.5YR) sandy clay loam pre-Mazama soil with many gravel to cobble-sized clasts, buried at most of the study sites, that formed in well-weathered basaltic and andesitic upland lava flows, colluvium from these residual soils, and some glacial outwash (Table 2.2). Undifferentiated pre-Mazama sediments of Unit I underlie high bench surfaces, some of which are probable pre-Mazama strath surfaces (S. Personius, personal communication, 1999). At some locations (e.g. Dry Creek and Lough Terrace) the soil formed in Unit I sediments appears to be quite old, Pleistocene or older, based on thick weathering rinds on cobbles and a 70 cm-thick Bt horizon. Otherwise, it has only been identified as pre-Mazama.

Unit II. This unit consists of primary Mazama ash flow and ash cloud deposits that filled the upper reaches of the North Umpqua River gorge as far downstream as Apple Creek, with traces as far west as Island Campground (Figs. 2.2, 2.3, Table 2.2). The primary ash flow sediment is fine, compacted ash with 10–30 % pumiceous lapilli and bombs. The ash cloud deposits are similar in composition, but consist primarily of fine ash with only fine pumice lapilli (Sherrod, personal communication, 1991). At the base of the primary ash is a diagnostic airfall layer of yellowish pumice lapilli 2–5 cm thick that was deposited just prior to the eruption of the glowing avalanches (Fig. 2.3a). Microprobe analysis (Foit 1991) confirmed that the Unit II sediments originated from the Mazama eruption 7600 cal. yrs B.P. (Table 2.1; Zdanowicz et al. 1999; Adams 1990). Unit II varies in thickness from approximately 25 m at Copeland Creek at the upstream end of the study area to about 2 m at Dry Creek, and 1 m thick at Lough Terrace (Figs. 2.1, 2.4, Table 2.2). Large quantities of this primary ash remain stabilized along the upper reaches of the North Umpqua River above Copeland Creek. Pockets of primary ash have been found as far downstream as Apple Creek and Island Campground. Where exposed at the surface, a weak to moderately developed soil has formed in this primary ash, weathering it to a lighter, yellower (10YR 6/6 brownish yellow dry), low-density soil.

Unit III. This unit is made up of well-bedded fluvially deposited sand and gravel, the original source of which was probably colluvium as well as glacial

Fig. 2.3 **a** Contact between paleosol of *reddish* gravelly clay, *below*, and Mazama ash *above*. Between these is the diagnostic layer of pumice lapilli and fine tephra that was erupted before the ash cloud and glowing avalanches roared down the river gorge. This layer is indicative of primary tephra above the contact. **b** Excavation site on Panther Terrace above the highway at Lough Terrace. **c** Sediments exposed in the test pit on Panther Terrace at Lough Terrace show, from bottom to top, Unit I pre-Mazama paleosol, Unit II primary Mazama tephra, Unit IV reworked Mazama tephra, and Unit VI post-Mazama colluvium and slope-wash sediments at the surface

outwash carried downstream from near the headwaters of the river (Table 2.2). These sediments are interbedded with layers of fluvially reworked Mazama ash and pumice lapilli. Little field-measurable soil development was observed in this unit, with the exception of some thin illuvial coatings on gravels at the Copeland Creek site (Fig. 2.4). However, coarse sediments such as these weather into soils very slowly. This unit represents post-eruption aggradation of the North Umpqua River channel. The unit ranges in thickness from about 1 m at Copeland Creek to about 3.5 m at Susan Creek Campground (Figs. 2.1, 2.4). Radiocarbon ages on stratigraphic units above and below Unit III suggest that this unit was probably deposited beginning sometime after the eruption, and continuing possibly as late as before 4500 years BP at Susan Creek and 3000 years BP at Copeland Creek (Table 2.1). At Susan Creek, radiocarbon ages on charcoal within the redeposited ash layers confirm the ash as Mazama in origin (Table 2.1; Musil 1994).

Unit IV. This is reworked Mazama ash, well weathered and generally of a fine loamy sand texture, low density (Table 2.2). This sediment was probably deposited from after 6000 to about 4500 years B.P. based on radiocarbon and hydration ages on archaeological deposits (Table 2.1). Unit IV is 0.6–1.6 m thick at Copeland Creek (Fig. 2.4), thickens to greater than 5.5 m at Island Campground (Fig. 2.6), and thins to less than 1 m at Susan Creek (Figs. 2.1, 2.7).

Unit V. The more recent floodplain sediments of Unit V contain little or no Mazama ash or pumice (Table 2.2). The parent material is sandy sediment, with occasional clay nodules, that probably originated on the slopes of the watershed upstream. This unit was deposited sometime after 1000 years B.P., and possibly within the last 500 years. It ranges in thickness from 60 cm at Island Campground to about 70 cm at Susan Creek Campground (Figs. 2.1, 2.6, 2.7).

Unit VI. The clayey colluvium of Unit VI was deposited by slopewash and mass movements on higher bench surfaces and portions of the highest river terrace intermittently throughout the past 7600 years (Table 2.2). Thus its age is only defined as post-Mazama. In color and texture it is similar to Unit I sediments (Fig. 2.3).

2.4.2 Structure and Age of the Terraces

Traces of possible pre-Mazama fluvial terrace formation and three post-Mazama terraces were identified along the North Umpqua River from Copeland Creek to Susan Creek (Table 2.3). From highest to lowest, the post-Mazama terraces, are

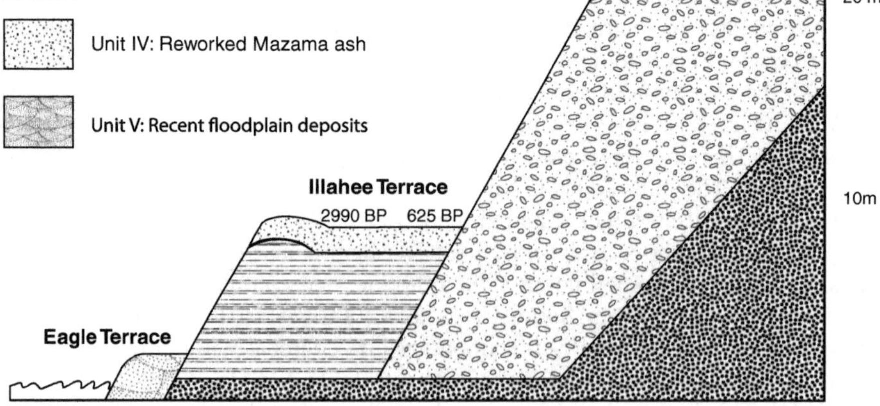

Fig. 2.4 Cross-sectional diagram of sediment units underlying terraces at Copeland Creek

referred to as Panther, Illahee, and Eagle Terraces. Figure 2.4 through 2.7 show the stratigraphy and structure of the terraces, from upstream to downstream, at Copeland Creek, Dry Creek, Island Campground, and Susan Creek Campground.

Pre-Mazama Terraces. Within the study area there are a few traces of possible pre-Mazama terraces exposed. These include the pre-Mazama (strath?) benches at Dry Creek (Fig. 2.5) and Apple Creek Bench (35DO418, 0.5 km downstream from Apple Creek Campground, Fig. 2.1) that appear to be covered with colluvium, and a narrow fragment of fill terrace observed up against the north wall of the river gorge at Susan Creek Campground (Fig. 2.7). Moreover, approximately 70 cm beneath the Eagle Terrace surface at both Island and Susan Creek Campgrounds is an apparent pre-Mazama paleosol (Figs. 2.6, 2.7).

Personius (1993; Personius et al. 1993) documented a fluvial terrace along the lower 200 km of the Umpqua River in the Coast Ranges. Radiocarbon ages on charcoal within the terrace deposits place the age of this terrace as early Holocene, 7–10 ka. Regional distribution and consistent age suggest that formation of this terrace was associated with a period of aggradation resulting from climate change during the Pleistocene-Holocene transition (Personius 1993). Further study of the pre-Mazama surfaces along the North Umpqua may reveal a temporal and genetic relationship with this Pleistocene-Holocene transition terrace of the lower Umpqua River.

Panther Terrace. Panther Terrace is the highest of the three terraces documented in this study, at 15–25 m above the modern channel, and represents a fill

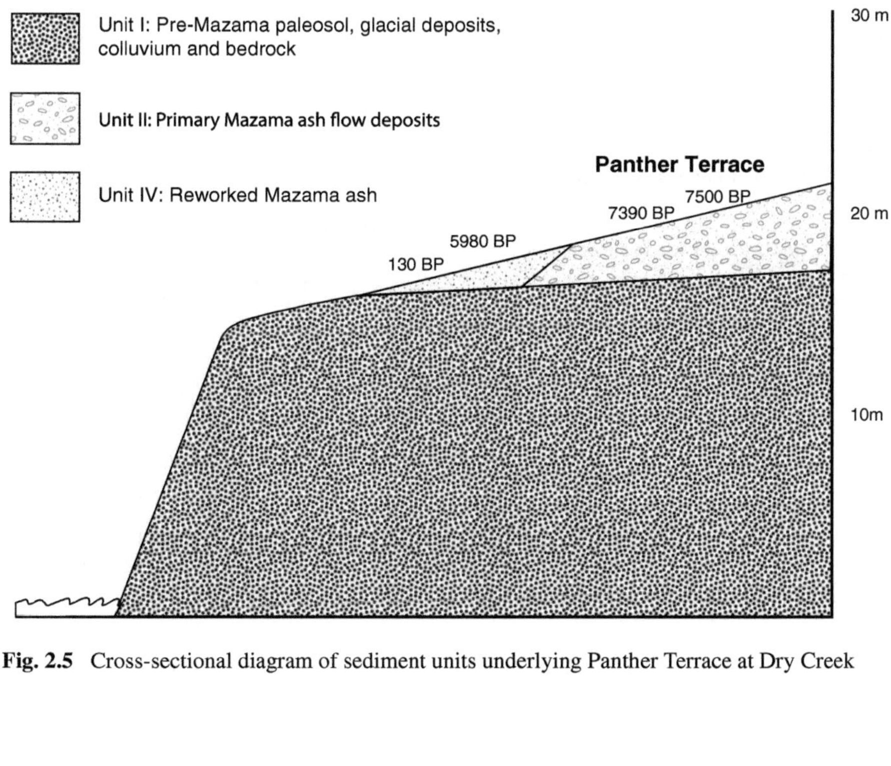

Fig. 2.5 Cross-sectional diagram of sediment units underlying Panther Terrace at Dry Creek

terrace that formed from the glowing avalanche sediments that flowed down the river gorge during the Mazama eruption 7627 ± 150 cal. yrs B.P. (Zdanowicz et al. 1999; Table 2.3, Figs. 2.2, 2.3, 2.4, 2.5, 2.6). At the upstream sites, Copeland Creek and Dry Creek, the Panther Terrace surface is underlain by Unit II primary ash flow and ash cloud deposits, overlying the Unit I paleosol (Tables 2.2, 2.3; Figs. 2.4, 2.5).

At Dry Creek, at the upslope end of the bench, 17.5 m above the modern channel, 2 m of primary ash (Unit II) overlies the pre-Mazama soil (Unit I) (Table 2.2; Fig. 2.5). Downslope, closer to the river at 15 m above the channel, the surface is mantled by a mix of primary ash and areas of disturbed and partially reworked ash. The latter was the location of Jenkins and Churchill's 5980 cal. yrs B.P. age (5220 ± 80 BP; Beta 31832; ETH 5523) on charcoal from the pre-Mazama paleosol (Jenkins and Churchill 1989; Table 2.1). This part of the terrace is composed of reworked Mazama deposits (Unit IV), likely disturbed when floodwaters reached midway up the bench and began to erode and rework the primary ash deposits. This would have occurred during channel aggradation, when Unit III was being emplaced and the channel bed was much higher than present (see Illahee Terrace, below). Thus, we believe the radiocarbon age of Jenkins and Churchill provides an approximate age on a period of erosion and reworking of some of the Mazama tephra deposits on a lower level at this site.

At Island Campground downstream (Fig. 2.6), Panther Terrace is underlain by intercalated Unit IV reworked Mazama ash and Unit VI clayey colluvium. Here,

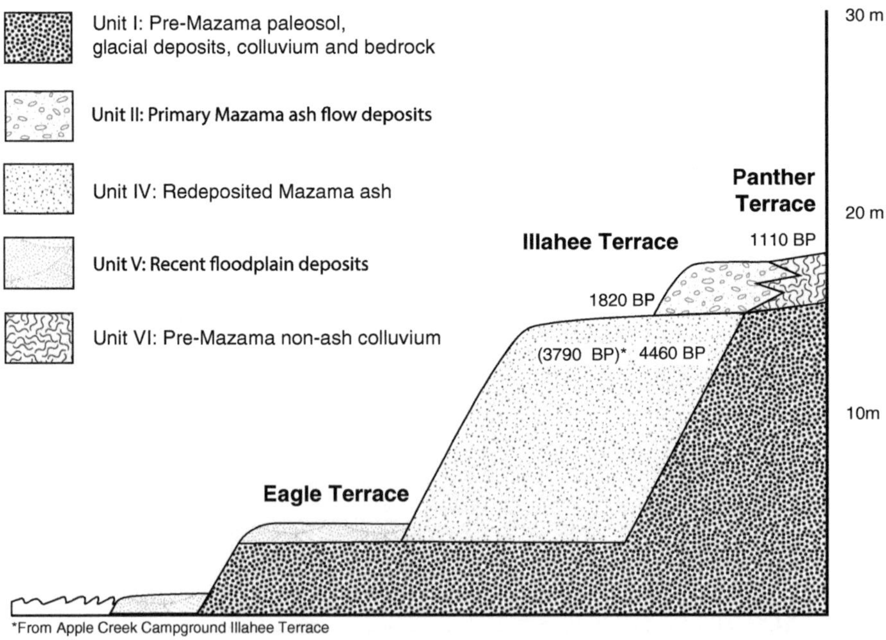

Fig. 2.6 Cross-sectional diagram of sediment units underlying terraces at Island Campground

beneath the Unit IV reworked ash is an 18-cm-thick layer of fine gray ash overlying a thin layer of pumice lapilli that appears to be Unit II primary ash flow deposits and the diagnostic air fall layer. If this interpretation is correct, Island Campground is the farthest downstream that primary ash flow deposits have been found. Beneath this primary ash is the Unit I paleosol. This combination of Unit II and Unit IV sediments could have been disturbed during the same flooding event that reworked the lower Dry Creek primary ash deposits, around 6000 years B.P (Figs. 2.5, 2.6).

Illahee Terrace. Illahee Terrace formed at 10.5–14 m above the modern channel, and is underlain by a sequence of Unit III and Unit IV post-Mazama alluvium (Tables 2.2, 2.3; Figs. 2.4, 2.6, 2.7).

Immediately after the Mazama eruption, and continuing intermittently for possibly as long as 3000 years, the North Umpqua River began to adjust its channel to flush the huge volume of pyroclastic debris in its channel downstream. The river reworked and transported both Mazama ash deposits and river and glacial gravels from the headwaters downstream, depositing them as the Unit III sediments that underlie the Illahee Terrace. We believe this represents the level of the active channel during that time, nearly 10 m above the modern channel at Copeland Creek and up to 13 m above the present channel at Susan Creek, 39 km downstream (Fig. 2.1). These well-bedded layers of ash and gravel were subsequently buried by the Unit IV overbank deposits of reworked Mazama ash throughout the length of the study area. Upstream at Copeland Creek, the river cut into the primary ash flow (Unit II)

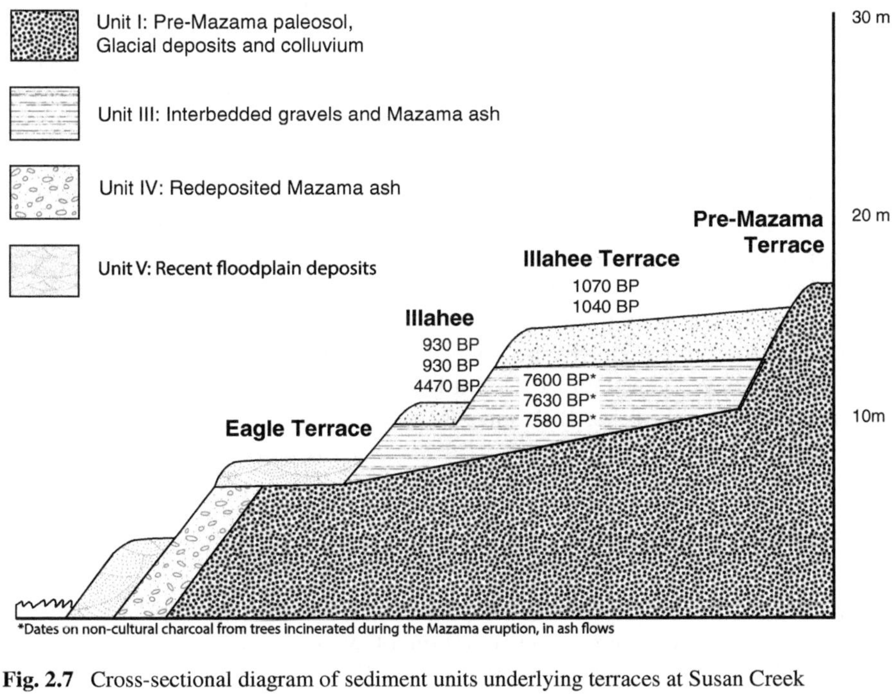

Fig. 2.7 Cross-sectional diagram of sediment units underlying terraces at Susan Creek

sediments and laid down the Unit III gravels and Unit IV overbank deposits on top of this erosional surface (Fig. 2.4, Tables 2.2, 2.3). At Susan Creek downstream, the Unit III gravels continued deposition until before 4500 cal. yrs B.P. (Fig. 2.7). Deposition on the Illahee Terrace ceased by around 4500 cal. yrs B.P. (Table 2.3).

The lower bench at Dry Creek, 12 m above the modern channel, is at the appropriate elevation corresponding with the Illahee Terrace, but has a different geomorphic origin (Fig. 2.5). Here the surface has been scoured of all traces of Mazama ash, leaving only the old, highly weathered cobbly clay paleosol in place. This surface has clearly been swept by past floodwaters rather than being formed by lateral migration of the channel; the bench slopes toward the river too steeply (12 % slope) to be a strath terrace cut by the channel. Moreover, the underlying sediments are poorly sorted cobbly clay sediments with common subangular boulders consistent with colluvium, probably mass movement deposits.

Eagle Terrace. At 5–8 m above the modern channel, the Eagle Terrace is underlain by late Holocene Unit V sediments, consisting of dark yellowish brown (10YR 3/4) very fine sand alluvium containing virtually no Mazama ash (Tables 2.1, 2.2, 2.3; Figs. 2.6, 2.7). These Unit V sediments may have begun deposition as early as 3000 years B.P. based on a single radiocarbon age at the North Bank site downstream from Glide (Isaac Barner, personal communication, 1991). The river may have begun aggrading at upstream sites by about 1100 years B.P. (Table 2.1). Eagle Terrace may still receive deposits from rare, high-level flood events, such as the December 1964 flood. At the Island Campground location of the Eagle Terrace, 60 cm of young Unit V sediments bury a moderately developed sandy clay loam paleosol (10YR 3/3) with no trace of Mazama ash, probably associated with Unit I pre-Mazama sediments (Fig. 2.6). The paleosol is characterized by a Bt horizon with moderate to strong medium subangular blocky structure. A similar profile is seen on the Eagle Terrace at Susan Creek Campground (Fig. 2.7). It is possible that this represents an older, pre-Mazama fill terrace surface that was buried by Mazama deposits and later exhumed. (The clay would be much more resistant to fluvial erosion than the light, sandy volcanic ash deposits.) Along the Eagle Terrace cutbank at Susan Creek Campground, the presence of an inset of Mazama ash may represent a remnant of Illahee Terrace-age floodplain deposits or it may be simply an old dump of ash excavated during road construction, as seen also on the lower bench at Dry Creek. With this exception, the general lack of Mazama ash observed in deposits underlying Eagle Terrace indicates that the Mazama ashflow deposits in the watershed have been stabilized for at least the past 1000 years or so, and possibly as early as 4500 cal. yrs B.P., since the abandonment of the Illahee Terrace.

2.5 Human Occupation

Archaeological remains indicate that people were living or camping along the North Umpqua River prior to the eruption of Mount Mazama. In the present study, pre-Mazama occupations were found at Dry Creek and Lough Terrace at the Panther Terrace level (Table 2.1, Fig. 2.3b). Obsidian hydration ages from one of the Panther Terrace components at the Apple Creek Campground and a radiocarbon date from the Narrows site (O'Neill 1989, 1991, 2002) suggest that people returned to occupy the banks of the river within a few hundred years after the eruption. Radiocarbon ages indicate that people occupied the Illahee Terrace during a possibly stable period in its development, between about 4500 and 3800 cal. yrs B.P., and also later around 1100–900 cal. yrs B.P. It is conceivable that between these two periods, from about 3800–1100 BP, cultural remains from that time could have been removed by river flooding, but no evidence of erosion from this surface has been noted within the sediment record. Given the obsidian hydration data from Island Campground, Lough Terrace, and Dry Creek archaeological deposits (Table 2.1), it is possible that occupation shifted to the more stable Panther Terrace. With the exception of the date of 3000 years B.P. from within Eagle sediments downstream from Glide (Isaac Barner, personal communication, 1991), there are also no data at present on human occupation of the Eagle Terrace surface. This may indicate that Eagle Terrace has been the active floodplain up until quite recently. Most importantly, however, there is insufficient evidence to answer definitively the question of whether the temporal groupings of cultural features along the river represent periods of stability in the river system or simply variations in human occupancy. Reasons for possible human presence/absence are not explored in this paper.

2.5.1 Prior Geoarchaeological Studies on the North Umpqua River

Studies of a number of other archaeological sites along the North Umpqua River have contributed to understanding the timing and formation of the three North Umpqua post-Mazama terraces. At Medicine Creek, upstream of Copeland Creek, Snyder (1981) encountered ash flow tephra deposits up to 2 m thick overlying a thin layer of airfall pumice. Snyder interpreted these sediments as primary and secondary Mazama tephra deposits overlying a reddish brown paleosol. The site was situated on a series of benches formed by mass movement deposits, the lowest of which was over 36 m above the river channel. This is 18 m higher than the highest alluvial terrace observed downstream. Snyder documented a pre-Mazama cultural component beneath the primary Mazama tephra at Medicine Creek (Snyder 1981).

The work of Jenkins and Churchill (1989) describes four sedimentary strata at Dry Creek, (Fig. 2.5) including three Mazama tephra layers overlying a compact, rocky brown sandy loam paleosol (Jenkins and Churchill 1989, p. 17). An AMS radiocarbon date of 5220 $\pm$ 80 (Beta-31832, ETH-5523) on charcoal from the upper 10 cm of the paleosol provides a calibrated age of 5980 years B.P. The authors concluded that the ash at Dry Creek could not be primary ash from the climactic eruption of Mazama. Therefore, they argued, the age assigned to the pre-Mazama archaeological component at Medicine Creek and other sites in the region might be in error (Jenkins and Churchill 1989, p. 18). However, our investigations at Dry Creek indicate that the northern, upslope section of the site is covered by primary Mazama ash while downslope and closer to the river the paleosol is overlain by reworked Mazama deposits. Our investigation of the Dry Creek site near where the authors obtained their radiocarbon sample indicates that the Mazama ash had been disturbed and portions had been redeposited. There is now clear evidence at Medicine Creek (cf. Snyder 1981; Keyser and Carlson 1983) and at Dry Creek that much of the tephra at these sites is undisturbed primary ash flow and ash cloud deposits. Jenkins and Churchill (1989) also obtained an obsidian hydration age of 4510 year BP on a composite sample of obsidian specimens obtained in the reworked Mazama deposits overlying the paleosol. This fits well with three radiocarbon ages and a hydration age obtained on terrace deposits at four sites downstream (Tables 2.1, 2.3; Fig. 2.5).

About 8 km downstream from Dry Creek at Apple Creek Campground, (35DO265; Fig. 2.1), O'Neill (1991) investigated cultural features on one of the terraces. Although not described in terms of parent material, the Apple Creek Campground terraces appear to correspond in elevation above the channel with the Illahee and Eagle terraces. A radiometric age of 3790 cal. yrs B.P. was obtained on charcoal in stratum 2 at a depth of 70 cm, which may correspond with radiocarbon dates on Illahee Terrace deposits at a similar depth at Island Campground (4460 cal. yrs B.P.) and at Susan Creek Campground (4470 cal. yrs B.P.). The Apple Creek Campground date is supported by an obsidian hydration age of 3500 years B.P. on artifacts from within the same stratum. Other hydration ages within the underlying stratum 3 of the same profile, 110–140 cm below surface, cluster at 5850–6300 yrs B.P., are consistent with the Dry Creek age, and provide a maximum age on deposition of Unit IV (Tables 2.1, 2.3).

Bogus Creek Campground is about 14 km downstream from Apple Creek and 6.8 km downstream from the confluence of Steamboat Creek with the North Umpqua River (Fig 2.1). Mairs (1987) described the context of the archaeological site at Bogus Creek on a rock-defended alluvial terrace extending from about 8–12 m above the channel. Although he described a single terrace, highway construction and other historical human disturbance may mask a scarp dividing this into two terraces, Eagle and Illahee. The brief description of sediments Mairs encountered seems to correlate with profiles seen at Island Campground, 8.4 km upstream, and Susan Creek Campground, 10.6 km downstream. They include fluvially reworked pumice and ash, with basalt pebbles and river cobbles particularly in the upper 30 cm of the soils (Mairs 1987).

Susan's Picnic site, about 11 km downstream from Bogus Creek and 1.5 km downstream from Susan's Creek Campground, was studied by Musil and Minor (1991; Fig. 2.1). The authors described two alluvial terraces that appear to correspond with the Illahee Terrace at Susan Creek, at 11 m above the modern channel, and the Eagle Terrace, 7 m above the river. However, sediments from the excavations were not described in terms of parent material, and no radiometric ages were obtained, so correlations with upstream terraces are tentative.

2.6 Discussion

This study describes the formation and evidence of human occupation of three alluvial terraces that formed at intervals along the North Umpqua River after the eruption of Mt. Mazama, 7600 years ago.

2.6.1 Provisional Chronology

Our study questions involve the timing and mechanisms of terrace development and abandonment. We have used primarily radiocarbon ages on cultural materials, supplemented by obsidian hydration on cultural artifacts, to provide maximum and minimum times of formation and abandonment of each terrace surface (Tables 2.1, 2.3). We acknowledge that more age control is needed to answer with confidence the question of ages of terrace formation and abandonment. However, we have made some estimates based on the evidence at hand. The dates cluster around specific times, beginning with the Mt. Mazama eruption at 7600 cal. yrs B.P., and including ca. 6000, 4500, and 1100–900 cal. yrs B.P. We base the following chronology on the location and stratigraphic context of the dated materials.

We estimate that the Panther Terrace initially formed of primary Mazama ash flow deposits at 7600 cal. yrs B.P. The edges of Panther closer to the channel were probably disturbed by floods around 6000 cal. yrs B.P. We believe the Panther surface formed the level of the North Umpqua channel only during the glowing avalanche from Mt. Mazama, and afterward was reached by floodwaters rarely except for the 6000 cal. yrs B.P. event. Thus this surface was virtually abandoned immediately after deposition (Tables 2.1, 2.3).

The foundations for Illahee Terrace began aggrading shortly after the eruption with deposition of the Unit III gravels. These channel gravels were encountered at the upper and lower ends of the study area, at Copeland Creek and Susan Creek, but likely also underlie Unit IV reworked Mazama ash at Island and Apple Creek Campgrounds. They indicate that the level of the channel at that time was 10–13 m above where the modern channel is today. A calibrated age of 4470 years B.P. on charcoal in Unit IV sediments a few cm above these gravels at Susan Creek, and a mean obsidian hydration age of 3000 years B.P. on seven samples at the boundary between Unit III and Unit IV at Copeland Creek, suggest that this phase of channel aggradation was very gradual and could have lasted thousands of years (Table 2.1). It also suggests that the deposition of Unit III gravels was likely time transgressive, beginning downstream and gradually propagating upstream over as much as 4600 years, ending at Susan Creek by 4500 cal. yrs B.P. and at Copeland Creek by 3000 cal. yrs B.P. (Figs. 2.4, 2.7). If this was the case, the channel during this phase would have been quite broad, with numerous gravel bars; it would have encompassed much of the gorge bottom leaving little or no floodplain. More sampling and radiometric ages on these sediments are needed to validate this hypothesis.

In its maximum aggradational phase, Illahee Terrace reached nearly full formation with the rapid deposition of Unit IV reworked Mazama ash, and was abandoned after 4500 cal. yrs B.P. Dates clustering around 1000 cal. yrs B.P. were assayed on cultural materials encountered within the top 50–60 cm on Illahee and Panther terraces. One possibility is that an unusually large flood deposited a final cap of reworked ash on both terraces sometime around 1100 cal. yrs B.P. However, the uniformity of sediment characteristics and diffuse boundaries between strata suggest that natural forces such as bioturbation may be more likely mechanisms for burial of the cultural features rather than additional fluvial aggradation. Rapid deposition of Unit IV is indicated, possibly in just a few hundred years or less. Thus, virtually all of the Mazama ash in the system was probably stabilized by around 4500 cal. yrs B.P. (Tables 2.1, 2.3, Figs. 2.4, 2.5, 2.6, 2.7).

The sediments of Eagle Terrace may have begun aggradation as early as 3000 cal. yrs B.P. (Isaac Barner, personal communication, 1991). These covered a possible strath terrace that had been scoured of the easily eroded volcanic ash down to a pre-Mazama surface. We obtained no radiometric or obsidian hydration ages on Eagle terrace materials. Weak soil development in the Unit V sediments, however, indicates a young surface, probably abandoned within the past few hundred years. It may still receive flood deposits during rare large flood events.

2.6.2 Mechanisms for Terrace Formation

The question of why these terraces formed is complex. Fluvial surfaces aggrade if sediment supply in the river is too great for the river's normal discharge to flush it downstream. Sediment supply will increase if slope erosion or mass movements become more frequent, often as a result of a change in climate. Yet volcanic eruptions such as that of Mt. Mazama will also abruptly increase sediment supply in the channel if glowing avalanches of ash and pumice flow down into the drainages. Aggradation raises the level of the channel bed and floodplain. When slope erosion rates are low, as when sediments are stabilized by vegetation or flood events are infrequent during a stable climate regime, the sediment supply decreases. Then the river discharge increases in erosive power and may begin to cut through the unconsolidated sediments in its bed, thus abandoning the level of the aggraded floodplain and leaving a terrace surface behind. Tectonic subsidence or uplift can also change the level of the channel, causing aggradation or incision (Schumm 1977, 1979, 2005; Schumm and Brakenridge 1987).

Two factors point to climate as a greater influence than tectonic adjustment along the North Umpqua. First, studies along the lower reaches of the Umpqua River as it flows through the Coast Range indicate that Holocene uplift rates in the Cascades are 0.2 +_0.3 mm/yr (Personius 1995). This rate would account for only approximately 15 cm of uplift since the Mazama eruption, whereas the lowest North Umpqua terrace is 5–8 m above the present channel (Table 2.3). Moreover, unless uplift occurs episodically and very rapidly, timing of aggradation or incision along a river system would be complex and time transgressive, that is, occurring at different times at different reaches along the length of the river. On the other hand, relatively synchronous changes in a fluvial system would suggest a climate-based stimulus rather than local complex response. It must be acknowledged, however, that teasing out the degree of influence of different mechanisms for terrace development is problematic.

In the case of the North Umpqua River terraces, there appears to be some evidence of the presence of one or more terraces prior to the eruption of Mt. Mazama, possibly associated with the Pleistocene-Holocene transition. However, when the glowing avalanches descended into the North Umpqua River drainage, they filled the channel and lower levels of the gorge to 25–17 m deep within the study area, 50–75 km downstream. The ash flows blanketed the pre-existing topography of the river banks.

We initially hypothesized that the proximal cause of disequilibrium in the North Umpqua River system that led to terrace formation was the Mt. Mazama eruption. This appears to be the likely stimulus for the formation and abandonment of the Panther Terrace and the emplacement of the Unit III channel gravels. However, there is evidence that a more regional cause may have contributed to final aggradation and abandonment on the Illahee Terrace around 4500 cal. yrs B.P., and formation of Eagle Terrace, beginning before 3000 cal. yrs B.P.

The 4500 cal. yrs B.P. ages in the study area appear to constrain the final development of Illahee Terrace, suggesting that the Unit IV sediments were deposited quite rapidly, over no more than about 700 years. The Unit IV ash deposits are for the most part quite pure and homogenous, not mixed with a lot of other materials, which seems consistent with rapid deposition, possibly related to only one or two very large flood events (with sediment supply from one or more large mass movements upstream). The ash deposits become more mixed farther downstream at Susan Creek. Although we might expect that the ash should have been entrained and redeposited much more immediately after the eruption, it is possible that little ash moved after the eruption during the warm, dry climate of the Altithermal (also referred to as the Hypsithermal, ca. 9,000–5000 B.P.). A long period of channel aggradation may have followed the eruption. Then a change in the environment before 4500 cal. yrs B.P. could have triggered a destabilization of the ash upstream. An event such as climate fluctuation, earthquake, or both, could have triggered large mass movements of ash into the river.

The pollen record at Gold Lake Bog, in the high western Cascades about 50 km north of the river headwaters (Sea and Whitlock 1995) suggests that the climate of the Oregon Cascades became moister and cooler than before around 5100 cal. yrs B.P. A similar change was recorded at Indian Prairie, in the western Cascades farther north by 4500 cal. yrs B.P. (Sea and Whitlock 1995). This data is supported by the pollen record from Little Lake, in the Coast Range farther north. The early Holocene record there reflects a warm, dry Altithermal climate with more frequent fires, followed sometime after 5600 BP by a cooler and more moist climate (Worona and Whitlock 1995). Perhaps the changes by about 5100 B.P. provoked one or more great mass movements of tephra, collapsing into the river channel upstream during a great flood. Data at Gold Lake Bog indicate the development of a mature forest and less frequent fire disturbance in the past 2000 years, which would lead to lower sediment input in the fluvial systems (Sea and Whitlock 1995). This would be consistent with stabilization of the Mazama ash, and later abandonment of the Eagle terrace.

2.7 Summary and Conclusions

Three post-Mazama terraces were identified along the North Umpqua River, named the Panther, Illahee, and Eagle Terraces. We suggest that the Panther Terrace formed as result of glowing avalanches that raced down the North Umpqua drainage during the eruption of Mt. Mazama, 7600 cal. yrs B.P. The underpinnings of Illahee Terrace likely began to aggrade the channel almost immediately after the eruption, principally from channel gravels interbedded with reworked ashflow deposits. The bulk of reworked Mazama ash that built the upper part of Illahee Terrace may have been deposited quite rapidly before around 4500 cal. yrs B.P., in response to a shift in climate from warm, dry conditions to cooler, moister conditions. Subsequently, the Mazama ash became stabilized and Illahee terrace was abandoned. Eagle terrace was aggraded on a pre-Mazama strath surface by non-Mazama sediments beginning after 3000 cal. yrs B.P. We conclude that although the Mount Mazama eruption disrupted the equilibrium of the North Umpqua River and provided large quantities of sediment to the drainage, climate also may have played a role in the cut and fill sequences that formed Illahee, and possibly Eagle, terraces along the upper reaches of the river. Further study of the North Umpqua terraces and human occupations on these terraces, and comparison with other regional geomorphic systems, may clarify whether a regional terrace-generating mechanism has been operating in the western Cascade region during the Holocene, and whether this mechanism is climate-based or a result of some other local or regional influence such as tectonic uplift of the Cascades.

References

Adams J (1990) Paleoseismicity of the Cascadia Subduction Zone: evidence from turbidites off the Oregon-Washington margin. Tectonics 9(4):569–583

Bacon CR (1983) Eruptive history of Mount Mazama and Crater Lake caldera, Cascade Range, U.S.A. J Volcanol Geother Res 18:57–115

Baldwin EM (1981) Geology of Oregon, 3rd edn. Kendall-Hunt Publishing Company, Dubuque, p 170

Bettis EA III, Benn DW (1984) An archaeological and geomorphological survey in the central Des Moines River Valley, Iowa. Plains Anthropol 29:211–227

Brakenridge GR (1984) Alluvial stratigraphy and radiocarbon dating along the Duck River, Tennessee: implications regarding flood-plain origin. Geol Soc Am Bull 95:9–25

Bull WB (1991) Geomorphic responses to climatic change. Oxford University Press, New York, p 326

Connolly TJ (1991) The Standley Site (35DO182): investigations into the prehistory of Camas Valley, Southwest Oregon: University of Oregon Anthropological Papers, vol 43. Department of Anthropology and Oregon State Museum of Anthropology, University of Oregon, Eugene, Oregon, p 168

Foit FF Jr. (1991) Volcanic glass analysis, Appendix D. In: O'Neill BL, Connolly TJ, Freidel DE (eds) Streamside occupations in the North Umpqua River drainage before and after the eruption of Mount Mazama: Oregon State Museum of Anthropology Report 96-2. University of Oregon, Eugene, pp 477–482

Franklin J, Dyrness CT (1973) Natural vegetation of Oregon and Washington: U.S.D.A. Forest Service General Technical Report PNW-8

Gardner GD, Donahue J (1985) The Little Platte drainage, Missouri: a model for locating temporal surfaces in a fluvial environment. In: Stein JK, Farrand WR (eds) Archaeological sediments in context. University of Maine, Center for the Study of Early Man, Orono, pp 69–89

Gee GW, Or D (2002) Particle-size analysis. In: Dane JH, Topp GC (eds) Methods of soil analysis, Part 4, physical methods. Soil Science Society of America Book Series #5 Soil Science Society of America. Madison, pp 255–293

Holliday VT (1987) Geoarchaeology and late quaternary geomorphology of the middle South Platte River, Northeastern Colorado. Geoarchaeology 2:317–329

Jenkins PC, Churchill TE (1989) Archaeological investigation of the Dry Creek Site, 35DO401: Report to the Umpqua National Forest by Coastal Magnetic Search and Survey. Salem, Oregon, p 71

Keyser JD, Carlson V (1983) Boundary determination for the Medicine Creek Site (35DO161): Report on file at the Umpqua National Forest. Roseburg, Oregon

Mairs JW (1987) Bogus Creek Campground site—geomorphology, Appendix II. In: Winthrop KR (ed) Bogus Creek data recovery project (35DO278): Report to the Umpqua National Forest by Winthrop Associates Cultural Research, Ashland, Oregon, B1-B3

Musil RR (1994) The archaeology of Susan Creek campground: Report to the Roseburg District Bureau of Land Management by Heritage Research Associates. Eugene, Oregon, p 202

Musil RR, Minor R (1991) Archaeological testing at Susan's Picnic site (35DO458), Douglas County. Report to the Roseburg District Bureau of Land Management by Heritage Research Associates, Oregon. Eugene, Oregon, p 90

Musil RR, O'Neill BL (1997) Source and distribution of archaeological obsidian in the Umpqua River Basin of southwest Oregon. In: Oetting AC (ed) Contributions to the archaeology of Oregon, Association of Oregon Archaeologists, Occasional Papers No. 6, Eugene, Oregon, pp 123–162

MacLeod NS, Sherrod DR, Chitwood LA, Jensen RA (1995) Geologic map of Newberry Volcano, Deschutes, Klamath, and Lake Counties, Oregon. US Department of the Interior, US Geological Survey, US Printing Office

O'Neill BL (1989) Archaeological investigations at the Narrows and Martin Creek sites, Douglas County, Oregon: Two riverside sites in the Middle Umpqua Basin. Cultural Resources Series No. 4, U.S. Department of the Interior, Bureau of Land Management, Oregon State Office, Portland, Oregon, p 239

O'Neill BL (1991) Evaluation of six archaeological sites along the North Umpqua Highway, Douglas County: Steamboat Creek to Boulder Flat Section: Oregon State Museum of Anthropology Report No. 91-1, University of Oregon, Eugene, p 151

O'Neill BL (2002) Multiple hydration rates on pre-Mazama obsidian artifacts in the Umpqua drainage, Southwest Oregon. CAHO 27(1/2):6–13

O'Neill BL, Connolly TJ, Freidel DE (1996) Streamside occupations in the North Umpqua River drainage before and after the eruption of Mount Mazama: Oregon State Museum of Anthropology Report 96-2, University of Oregon, Eugene, p 391

Personius SF (1995) Late quaternary stream incision and uplift in the forearc of the Cascadia Subduction Zone, Western Oregon. J Geophys Res, vol 100, pp 20,193–20,210

Personius SF (1993) Age and origin of fluvial terraces in the Central Coast Range, Western Oregon. U.S. Geological Survey Bulletin 2038. USGPO, Washington, pp 56

Personius SF, Kelsey HM, Grabau PC (1993) Evidence for regional stream aggradation in the Central Oregon Coast Range during the Pleistocene-Holocene transition. Quatern Res 40:297–308

Ritter DF, Kochel RC, Miller JR (2011) Process geomorphology, 5th edn. Wm. C. Brown, Dubuque, pp 280–282

Schoeneberger PJ, Wysocki DA, Benham EC, Broderson WD (1998) Field book for describing and sampling soils. Natural Resources Conservation Service, USDA National Soil Survey Center, Lincoln

Schumm SA (1977) The fluvial system. Wiley, New York, p 338

Schumm SA (1979) Geomorphic thresholds: the concept and its applications. Prog Phys Geogr 4:485–515

Schumm SA (2005) River variability and complexity. Cambridge University Press, Cambridge

Schumm SA, Brakenridge GR (1987) River responses. In: Ruddiman WF, Wright HE Jr. (eds) North America and adjacent oceans during the last deglaciation, vol K-3. Geological Society of America, The Geology of North America, Boulder, Colorado, pp 221–240

Sea DS, Whitlock C (1995) Postglacial vegetation and climate of the Cascade Range, central Oregon. Quatern Res 43:370–381

Sherrod DR (1991) Geologic map of a part of the Cascade Range between latitudes 43°–44°. U.S. Geological Survey Miscellaneous Investigation Series, scale 1:125,000, Central Oregon

Snyder SL (1981) Medicine Creek pre- and post-Mazama Occupation in the Cascades. : Tebiwa 23:1–13

Stuiver, M. and Reimer, P.J. (2011) Radiocarbon calibration program Rev 6.0. Quaternary Isotope Lab, University of Washington, Seattle

U.S. Geological Survey, (accessed 12/5/2012) USGS 14316500 N Umpqua River Above Copeland Ck Nr Toketee Falls, OR, USGS Real-Time Water Data for the Nation, http://waterdata.usgs.gov/nwis/uv?

U.S. Geological Survey (2011) USGS Water Data Report 2011 14316500 N Umpqua River Above Copeland Creek, Near Toketee Falls, OR. Wdr.water.usgs.gov/wy2011/pdfs/14316500.2011.pdf

Wert SR et al (1977) Soil inventory of the Roseburg District: Bureau of Land Management. Roseburg, Oregon

Worona MA, Whitlock C (1995) Late quaternary vegetation and climate history near Little Lake, central Coast Range, Oregon. GSA Bulletin 107(7):867–876

Zdanowicz CM, Zielinski GA, Germani MS (1999) Mount Mazama eruption calendrical age verified and atmospheric impact assessed. Geology 27(7):621–624

Chapter 3
Prehistoric Settlement Patterns and Optimal Maize Field Location in the Mt. Trumbull Region NW Arizona USA

Paul E. Buck and Donald E. Sabol

Abstract Optimal maize field locations possibly used by prehistoric agriculturalists ~A.D. 1100 in Mt. Trumbull area of northwest Arizona were modeled using remotely sensed data and ground based observations. Models were created by developing GIS "layers" using radiant surface temperature from satellite (ASTER, the Advanced Spacebourne Thermal Emission and Reflection Radiometer), digital orthophoto quad data, 10 m DEM data, and NRCS soil maps. Near surface air temperature and soil moisture were collected on the ground. Two distinct crop/habitation models are constructed: a "restrictive" model and a "fuzzy logic" model. The proximity of the optimal field locations is compared with the distribution of archaeological sites, specifically investigating whether one or two room structures with sparse artifact scatters (so called "field houses") are indeed preferentially associated with optimal field locations as some have suggested. Results indicate that "field houses" are not preferentially associated with the most optimal field locations. Rather the larger habitations sites consisting of multi-room or larger C shaped habitations are significantly more likely to be found within 200 m of an optimal field location. Common "field houses" may in fact have served other purposes, or could reflect more intensive use of sub-optimal locations later in the prehistoric sequence nearer the eventual abandonment of the area in late Pueblo II or early Pueblo III times. Better chronological control of occupations of all these sites will be useful in explaining this pattern.

P. E. Buck (✉)
Desert Research Institute, 755 E Flamingo Rd, Las Vegas, NV 89119, USA
e-mail: Paul.Buck@nsc.edu

D. E. Sabol
Desert Research Institute, 2215 Raggio Parkway, Reno, NV 89512, USA

M. Goman, *Human Environment Interactions – Volume 2*,
SpringerBriefs in Earth System Sciences, DOI: 10.1007/978-3-642-36880-6_3,

3.1 Introduction

The goal of this project is to identify agricultural fields—locations on the landscape where the Ancestral Puebloans (the Anasazi in archaeological literature) grew crops. The locations may have changed through time. Unlike modern farming techniques where large plots of land are used year after year with regular inputs of labor and fertilizers, ancient Anasazi "farms" were more likely to have been ephemeral locations on the landscape during the period of Basketmaker II to Early Pueblo III, roughly A.D. 1 to A.D. 1200 (Lyneis 1995). Factors contributing to the productivity of places may have included water availability, soil, aspect, slope, ground cover and others. Precipitation patterns, and seasonality of rainfall, may have been determining factors.

The Anasazi of the Uinkaret Plateau (see Fig. 3.1), and indeed of most of the Colorado Plateau, are thought to have relied on maize agriculture (to a greater or lesser extent, see Cordell 2009 for example). Since there are few perennial streams, agricultural populations in this area were forced to rely only on rainfall or dry farming. We recognize (thanks to Steadman Upham 1984) that populations could theoretically and likely did move in and out of specialized agriculture depending on resource nature and abundance—there is no a priori assumption that all inhabitants of the study area were full time agriculturalists at any given time. To minimize the likelihood of a catastrophic crop failure, maize was likely grown in small plots, probably in scattered locations, some close to villages, some farther away perhaps even in different hydrologic basins or catchments. To accomplish this, each

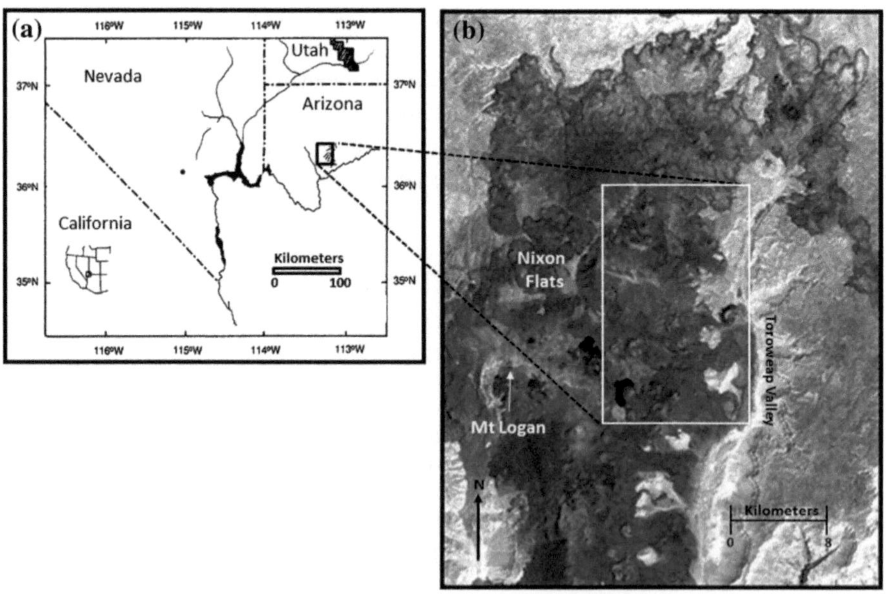

Fig. 3.1 Location map. Mt. Trumbull is part of the Colorado Plateau located in northwest Arizona just north of the Colorado river

"co-residential unit" (Lyneis 1995) or kin group needed to establish fields and possibly an associated "field house" (an archaeological unit consisting of one or two simple rooms with few associated artifacts; see extended discussion below) dispersed around a permanent habitation. This meant that in any given area one might expect to find field houses belonging to unrelated or distantly related kin groups, and not have the catchment monopolized solely by a single group. It if were, the chances of a family or kin group crop failure would be higher. If productivity of smaller catchments is highly variable over space and time, then it makes sense for individual villages/residential sites' households/communities to disperse their "ownership claim" (i.e., field houses, Kohler 1992) across as many of these catchments as possible. If one field fails, that failure does not jeopardize the entire crop.

Why did the ancient Puebloans of Mt. Trumbull, on the western edge of the Colorado Plateau, place their settlements where they did? Were they tied to maize fields for at least part of the year? Did settlement location change overtime, and if so, why?

There is currently no good regional scale method for assessing potential for maize agriculture independent of simply counting the number of archaeological sites thought to be related to agricultural activity (viz., check dams, cobble mulch fields, and "field houses") in an area and arguing that fields are nearby. One measure of reliance on agriculture then might be examined through understanding settlement location. If the Anasazi of Mt. Trumbulll relied on maize agriculture, then it might be expected that settlement choices were at least partially conditioned by their proximity to suitable agricultural land.

Ethnographic accounts of traditional agricultural practices in the Southwest suggest small areas and widely spaced crops within each area. Suitable areas may have been no more than a few hectares in size, with maize planted in small hills several meters apart. Because of the unpredictability of adequate precipitation or groundwater in any one given area, it may have taken several (or dozens) of small farm plots dispersed over a fairly large area to support a household—perhaps within a few kilometers of a larger pueblo which functioned as the main habitation.

In many areas of the Southwest, a number of features have been found which indicate use of an area as agricultural fields. Rock piles, cleared areas, gridded areas, linear rock alignments across the slope, and others have all been interpreted as agricultural technology designed to maximize yields of fields (Cordell 2009, pp. 285–296; Herhahn 1995). These features are located on the field or very close by. These kinds of features have rarely been found in the Mt. Trumbull area; instead certain site types have been used (Moffit and Chang 1978) as a (tautological) means of identifying possible agricultural fields. "Field houses" are assigned that name because of their presumed spatial proximity to fields. These are often one or two room structures distant from the main habitation pueblo with few artifacts on the surface. Actual "fields" have not been identified near these "field houses"; the mere naming of a site as a field house is an assertion that agricultural fields were close by. No independent verification of this is usually provided.

The term "field house" has a long history in southwestern archaeological literature. Possible explanations for one or two room structures with few artifacts located generally some distance from large residential Pueblos include: (1) as "trial" settlements or "starter" villages (what Ellis 1978, p. 60 called "an infant pueblo"); (2) as property claims or markers of territory (e.g., Preucel 1988); and (3) as agricultural monitoring stations ("field houses") or farmsteads (Haury 1956). The identification and purposes of those sites referred to as "field houses" is often problematic. These have been claimed to be associated with "good" soils, although it is admitted that "…no actual field locations, other than those presumably adjacent to field houses, can be recognized in the local archaeological record" (Kohler 1992, p. 626).

Although some have argued that they represent intensification of agriculture and that people had to go farther and farther from villages to protect their increasingly distant crops, it is clear at Mt. Trumbull that some field houses are quite close to (presumed) residences of the owners of the fields and their crops. Smaller structural sites are more numerous than larger C-shaped pueblos which may have 15 or more rooms; these are the largest structural sites in the study area, and may consist of up to 20 or more rooms of various sizes generally organized in a "C" or even an enclosed oval pattern. These large sites may show evidence of a surface depression indicating a pit house, have a range of different rooms suggesting both habitation and food storage, and often have 50,000 or more artifacts on the surface including clear midden areas. Data from Mt. Trumbull shows that for every C-or L- shaped pueblo there are five or six other structural sites including some single room "field houses", others with two small rooms, and others with three to eight rooms.

We do not know if the smaller rooms and field houses nearest the larger pueblos are "owned" or established by the co-residential group living at the pueblo. Occupations of these smaller sites may be from more distant pueblos in an effort to minimize risk of crop failure by dispersing their fields over a wider area.

The central problem is this—how can one independently identify optimum locations for productive agricultural sites where there are few signs remaining of alteration or modification of a landscape? The identification and subsequent estimation of size of fields is essential for reconstructing carrying capacity—for comparing demographic and population trends in one area with another. The amount of acreage used may indicate intensification of agriculture, environmental stress, or changes in precipitation patterns. We must learn about the agricultural carrying capacity of a given region if we wish to understand population dynamics—growth, aggregation, dispersion, and local abandonment (Rhode 1995; Leonard 1989).

3.1.1 Key Objectives of the Work

This project uses data from a variety of ground-based and remote sensing platforms and combines them in a GIS to look for patterns in ancient use, surface characteristics, and composition to predict possible ancient maize agricultural fields. The primary application of remote sensing for this project is to assist in identifying areas of suitable temperature and soil moisture which might have enhanced conditions for productive maize agriculture.

Our objectives are to: (1) link remotely sensed data of the Mt. Trumbull region of northwest Arizona (an area of ca. 40,000 ha) with ground based observations of soils, vegetation, and hydrology and an understanding of maize ecology to identify optimal places on the landscape for ancient Anasazi agriculture; (2) link known archaeological sites of various types (especially the so-called "field houses") from a subset of the study area with these optimal field areas as a "training set" and then construct a predictive model for other portions of the study area; (3) test this model using data from a previously archaeologically unknown portion of the Mt. Trumbull region to validate (or not) the linkage between optimal field areas against archaeological survey data; (4) finally, examine whether or not field houses (or any other site type) are in fact preferentially associated with these optimal areas.

We use remotely sensed data to develop a model of Anasazi agriculture in the Mt. Trumbull area of the Uinkaret Plateau, a little studied part of the American Southwest Colorado Plateau. This model can also assist federal land managers in targeting areas for archaeological survey, where less than 3 % of the Grand Canyon Parashant National Monument has been inspected and federal agency funding is stagnant.

3.2 Background

The study area (Fig. 3.1) is part of the Uinkaret Plateau, itself part of the Colorado Plateau. It is a semiarid region ranging from 1524–2438 masl in elevation just north of the Colorado River in Arizona. Below we focus mostly on aspects of climate relevant to prehistoric agriculture, especially successful cultivation of maize, thought to have been a staple food in the Anasazi diet. In this part of the Uinkaret Plateau it is thought that cultivation was based on dry-farming (no irrigation).

3.2.1 Modern Climate

The two main climatic factors controlling cultivation for maize were likely precipitation (and other sources of water such as springs) and number of frost free days, or minimum temperature. There are four meteorological stations in the general area with relatively long climate/weather records; these are from west to east Mount Trumbull Schoolhouse (in the now-abandoned town sometimes called "Mount Trumbull" or "Bundyville" but not to be confused with the actual mountain called "Mount Trumbull" some eight miles east); Nixon Flats; Mount Logan; and Tuweep, a ranger station near Toroweap on the north rim of the Grand Canyon. Figure 3.2 shows the seasonality of precipitation for each of the stations. There is a great deal of relief and topography in the area due to Mt. Trumbull and Mount Logan—it is likely that considerable variety in precipitation and temperature exists in this study area that is not faithfully captured by these four stations. Average annual precipitation ranges from somewhat less than 300 mm to almost 460 mm at Nixon Flats. Mount Logan, located almost 213 m higher than Nixon Flats, receives on average 127 mm of precipitation less per year. In fact Mount Logan receives almost the same precipitation as the two stations far lower in altitude at Tuweep (1500 m) and Mount

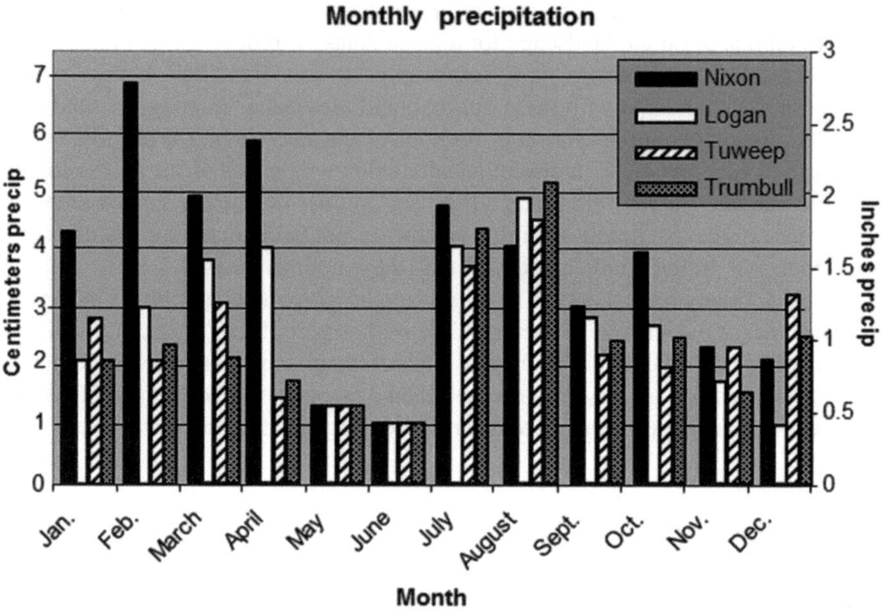

Fig. 3.2 Seasonality of precipitation as shown from Mt. Logan and Nixon Flats meteorological stations

Trumbull schoolhouse (1707 m). This is seen in Fig. 3.2. This figure also shows that the winter/spring precipitation is the main distinguishing feature of Nixon Flats. This is mostly likely in the form of snow since the average temperatures during periods of winter/spring precipitation are below freezing.

Of course, there is a clear seasonal pattern to the precipitation—it is a monsoonal pattern with about half of the precipitation coming July, August, and September. Interestingly, the amounts of rainfall in all four stations during these months are very similar. May and June are the driest months of the year. The amount of precipitation is also highly variable—in the 20 year record of Nixon Flats, 1 year (1993) had 1040 mm of precipitation, but 2009 had only 200 mm.

3.2.2 Geology

The study area is part of the Uinkaret Plateau, a sub-region of the Colorado Plateau. It has been described most recently and completely by Billingsley and Hamblin (2001) whose descriptions include a detailed surficial geological map of the four USGS quads that coincide with the archaeological study area. Much of the description below is from that report and the accompanying map. The study area is dominated by Tertiary and Quaternary volcanic deposits and surficial deposits. An estimated 2/3 of the entire study area is covered by rocks of these ages; the remainder is exposed Paleozoic and Mesozoic sedimentary rocks found to the east and west especially as exposures in Toroweap Valley.

The most common units are Quaternary lava flows and pyroclastic deposits of less than 1 million years in age. The most recent flow is the Little Springs flow dated to about 1,000 years ago. These lavas flows are all comprised of alkali-olivine basalts (Billingsley and Hamblin 2001). Flows are covered by thin soils in many places, and are estimated to be perhaps 100,000 years old. There are an estimated 154 Quaternary cinder cones in the study area. These are pyroclastic deposits created at approximately the same time as the flows, and can be from several meters thick up to 100 s of meters thick at the cones themselves. These pyroclastic deposits are found as drapes or thin veneers covering thousands of hectares in the study area. The alkali-olivine basalts are the source of one of the most characteristic materials found in prehistoric pottery from the Mt. Trumbull area; the olivine phenocrysts used as tempering (non-plastic) materials added to clay for pottery manufacture. It is the presence of this olivine that allows pottery to be traced back to the Mt. Trumbull area.

Other kinds of deposits are also present in smaller amounts. For example, there is a wide zone of talus around Mt. Trumbull and Mt. Logan, resulting from the collapse of the basalt cap on top of older sedimentary rocks which has then draped the steep slopes of these two mountains. But these are quite steep and unsuited for agriculture, much less human habitation.

The other relatively important unit is composed of Holocene deposits, including stream channel alluvium, floodplain, terrace gravel deposits, alluvial fan deposits, colluviums and valley fill. These deposits are found in small quantities immediately in the Mt Trumbull area, but cover much more area to the east where they are widespread in Toroweap Valley (see Fig. 3.2). Many small exposures of Holocene deposits are too small to be mapped on the geological map of Billingsley and Hamblin (2001). These small areas, often no more than a few meters in length and often found in the low lying valleys between the cinder cones, may have been focal points for agriculture.

Another aspect of the geology may be just as important as the rock types—this is topography created by the abundant cinder cones. The relief created by these features, and the relative proximity of variable relief features at any given place on the Uinkaret Plateau, mean that a variety of slopes, aspects, and landscape relief was available to an ancient farmer living near Mt.Trumbull. It is only a matter of an hour walk to have a different exposure or slope on which crops could potentially have been grown. This creates a great variety of microclimatic zones in a relatively compact area. This topography may have created locally favorable accumulations of winter snow, or induced (through orographic effects) more precipitation in certain places.

3.2.3 Soils

Most of Mt. Trumbull area is covered by "well drained and somewhat excessively drained soils formed in residuum and alluvium weathered from basalt, andesite, ash-flow tuff, cinders and related volcanic rock. They occur on plains, mesa tops and cinder cones" (Hendricks 1985, p. 126). TA1 soils are classified as Thermic

Arid soils and have mean annual soil temperature of 15–22 °C. TA1 series are shallow and moderately deep, moderately sloping to extremely steep, gravelly cobly and stone, moderately coarse to moderately fine textured soils developed in colluviums and on residual materials and as limestone, sandstone, and shale bedrock. They are found in the Toroweap Valley as valley fill. More recent information about the soils is available from Jorgensen (2004) or on the web soil survey homepage (http://websoilsurvey.nrcs.usda.gov). This recent soil survey is a very detailed report on all the soils in the Mt. Trumbull study area, including information about soil setting, landform position, and depth to water table, characteristics of the parent material, water holding capacity, infiltration capacity and other important characteristics.

3.2.4 Flora and Fauna

Most of the study area is comprised of pinyon-juniper (*Pinus-Juniperus*) woodland and at higher elevations of Mt. Trumbull and Mt. Logan montane conifer forest. These areas are characterized by ponderosa pine (*Pinus ponderosa*) forests, with an understory of Gambel oak (*Quercus gambelli*), Arizona madrone (*Arbutus arizonica*), big-tooth maple (*Acer grandidetatum*), and quaking aspen (*Populous tremuloides*). Many stands in the area are open or park like (especially where natural fires have not been restrained. The Mt. Trumbull/Mt. Logan area has approximately 7600 ha of ponderosa pine forest (Arizona Strip Final Environmental Impact Statement 2006). The species most commonly associated with ponderosa pine is Gambel oak. Small clumps of quaking aspen may also grow in the general area, often near a meadow. Other species include New Mexican locust (*Rubinia neomexicana*) and serviceberry (*Amelanchier utahensis*), both usually as shrubs or small trees. The understory of more open stands supports abundant grasses and forbs. Shrubs present include those from adjoining communities along with scattered individuals of mountain snowberry *Symphoricarpus opreophilas*), Oregon grape (*Mahonioa aquifolia*), Utah juniper (*Juniperus osteosperma*), and Oregon boxwood (*Pachistima myrsinities*). Several species of wildlife are dependent upon ponderosa pine, including Kaibab squirrels (*Scirus aberti kaibabensis*), goshawks (*Accipiter* sp.), and Merriam's turkey (*Meleagris galloparo*).

Most areas below about 2134 m in the study area are covered with pinyon-juniper woodland. Pinyon and juniper are the dominant tree species of this zone in northern Arizona. The species of pinyon most often present is the common pinyon (two-leaf pinyon or Colorado pinyon, *Pinus edulis*), with single leaf pinyon (*Pinus monophylla*) occurring occasionally. Utah juniper is the most common juniper present, with one-seed juniper occasionally found. The under stories of pinyon-juniper and dense mature juniper woodlands are very species-poor, containing only widely scattered shrubs, forbs, and small clumps of grass. Grasses are the most common under story component. Predominant (or formerly predominant) grasses include grama (*Bouteloura* ssp.), Arizona fescue (*Festuca arizonica*), prairie June grass (*Koeleria macrantha*), Indian rice grass (*Oryzyopsis hymeoides*), needle grass (*Stipa* ssp.) dropseed (*Sporotulus* sp.), and squirrel tail (*Hordeum jubatum*). Shrubs may include sagebrush (*Artemesia* sp.), cliff rose (*Purshia* sp.), serviceberry, rabbit brush (*Chrysothamnus nauseosus*), shad scale (*Atriplex* sp.), and winter fat (*Krascheninnikovia lanata*). Under story plants are most common along the edges of the zone. Bare ground is very common. Utah juniper is a climax species in a number of pinyon juniper, sagebrush, grassland, and shrub-steppe communities. The natural fire regime of these pinyon-juniper areas ranges from frequent to infrequent fire return intervals of between 30–100 years apart with mixed to local stand replacement fire severity.

3.2.5 Archaeological Background

This part of northwestern Arizona has been little studied by archaeologists. It is part of the western fringes of the Anasazi people (Puebloans) of the American Southwest, centered generally on the "Four Corners" area, near the intersection of four western states: Arizona, Colorado, New Mexico, and Utah. With the exception of a few small surveys in the region, cultural resources management projects in the area did not begin until the mid-1970s (Lyneis 1995; Reid and Whittlesey 1997). At this time a transect random sample survey of about 800 ha of the study area was completed (Moffit and Chang 1978). Seventy two prehistoric sites were found, including a number of larger (~20 room) C-shaped pueblos and other habitations. Most were found in the pinyon juniper zone near Mt. Trumbull and Mt. Logan. A number of non-habitation sites were also noted, such as "field houses" and "artifact scatters." Since that time, the Bureau of Land Management (BLM, an agency of the US federal government that manages millions of hectares in the western US) has conducted many small scale surveys for compliance with section 106 of the National Historic Policy Act and recorded a total of perhaps 400 prehistoric archaeological sites in the study area (site files on record at BLM office, St. George Utah). Local and regional archaeologists had suspected for some time (Altschul and Fairley 1989) that prehistoric sites were abundant at Mt. Trumbull in comparison to sites further east and west.

There was little archaeological research per se conducted in the study area until Buck (n.d.) began work in 2001. Prior to this time many sites had been recorded by amateur groups especially the Utah Statewide Archaeological Society under the leadership of the Arizona Strip District Archaeologist Diana Hawks. These tended to be in areas proposed for chaining activities, roads, forest restoration treatment or other ground disturbing actions requiring archaeological survey by NHPA or National Environmental Policy Act. Chaining, a practice common in the western US to create grassland suitable for cattle, flattened many Anasazi architectural sites in the area since it involves pulling a very large heavy chain linked by two Caterpillar tractors across the landscape, knocking down small trees and shrubs. Areas are then seeded with grass edible to cattle. Most sites are Pueblo II in age (i.e. estimated about A.D. 1000–1200) and found preferentially in the pinyon-juniper zone, in part because sites in the slightly higher elevation ponderosa pine forests are obscured by thick forest duff. Earlier sites are seemingly underrepresented, but some possible storage cysts and lithic scatters attributable to the earlier Basket Maker or Archaic period are known. No substantial excavations have been conducted in the study area, although minor testing has been conducted at a number of sites in the study area and reported radiocarbon dates ranging from A.D. 600 to 1200 (Buck and Sakai 2005).

In 2001 a long-term program of archaeology research was started in the study area by the one of the authors (Buck 2002). This has been part of the joint Nevada State College/Desert Research Institute archaeological field school conducted in the area in 2001 and 2003–2008. The concept of adaptive diversity (the nomad-agriculturalist continuum) as described by Upham (1984) provides a useful heuristic tool for examining the archaeological record of the western District of the Virgin Ancestral Puebloan. It focuses attention on the entire archaeological record, not merely the most obvious portion of it. It requires attention to chronology, and identifies potential linkages between larger and smaller sites as an important topic of research. A number of (mostly habitation) sites have been tested (Buck 2002; Buck et al. 2004; Buck 2005, 2006). In addition, about 567 ha have been intensively surveyed to date (Fig. 3.2, Buck 2005, 2006). Because of this kind of intensive survey, about 120 new sites have been recorded in a portion of the study area, where a reported site density of less than 1/5 this amount was known prior. Studies are being conducted of the lithics (Martin 2009) and ceramics from many of these sites (Sakai 2005, 2007, 2010, 2012).

3.2.6 Distance Between Field Houses and Fields: Expectations from Ethnography and Archaeology

How close might small structures (one or two rooms with few artifacts viz "field houses") need to be to optimal field locations to be considered "close"?? That is, if these small structure indeed functioned as field houses (to watch over fields) then is there some maximum distance away from fields that permits them to function in that capacity? A field house so far away from the field that animals and other predators cannot be seen or stopped is useless; it does no good to have a field house that is too far away to monitor a field. We hypothesize this maximum distance to have been ~100 m.

There is ethnographic, archaeological, and modern evidence to support this distance (Moore 1978; Mindeleff 1891; Russell 1978; Honeycutt 1995; Bradfield 1971; Kohler 1992; Lang 1995; Herhahn 1995). The ethnographic evidence comes from Hopi and Zuni areas where ethnographic and historic accounts chronicle small structures (sometimes called windbreaks or field houses) located near crops, primarily to monitor the crop and prevent predation by animals, or for weeding. The archaeological evidence below examines only those situations where there are clear agricultural features which likely truly denote places where maize was grown. Such features include rock piles, bordered fields, linear garden borders, mulch fields, check dams, and terracing. These features represent archaeological evidence of crops grown at that place. There is also an account of contemporary Anglo dry farming that suggests small structures very close to where crops are grown would have resulted in much higher yields.

The field houses are thought to have housed people (although perhaps also tools, crops, water) to watch over the crop and to protect it from predators. Of course the people in the larger habitations at Mt. Trumbull—the 12–20 room C-shaped pueblos and other somewhat smaller villages on little hills—could easily have watched over their crops within eyesight. The apparent proximity of "field houses" to the large sites might be more apparent than real—it could be that they were not used at the same time. Field houses might have been established by someone from another village after the nearest larger habitation structure had been abandoned, or before it was constructed.

Evidence from ethnography. A variety of Puebloan ethnographic sources from the Southwestern US suggest small structures (perishable and not) were habitually located in or close to fields. A field house "...may be situated within, on the edge or overlooking it [the field], in close visual proximity" implying a locus with unimpaired visibility and a distance to the field of "no more than a stone's throw" (Moore 1978, p. 10). Mindeleff (1891), p. 642 described contemporary Tusayan "farming shelters, temporary establishments occupied only during the farming season and abandoned on the approach of winter, but located directly on or overlooking the fields under cultivation" (emphasis added) when discussing Pueblo architecture in the New Mexico. Navajo agricultural fields were usually located a "short distance" from the field with ~100 m as the maximum distance for optimal location (Russell 1978, p. 38).

Maitland Bradfield (1971) lived among the Hopi in the 1960s and examined the factors that dictated their choice of field sites. He concludes that it is likely that most fields were less than 6.5 km from habitation villages, in part because of the extreme efforts of carrying so much crop back to villages for storage. He also believes that the danger of raids by enemies such as Navajo and Apaches caused fields to be closer rather than farther. Further "that distance from the fields has always been the limiting factor determining the size of villages"—so when the maximum productivity of the field area was reached the village split or fissioned into a second village.

Evidence from archaeology. Kohler (1992) argues that maize was such an important part of the Anasazi diet that monitoring of the growing crop was essential (1992), p. 620 and could not be left to chance. For the Dolores Archaeological Project (DAP), a multi year extensive excavation effort in Southwest Colorado (reported in Kohler 1992), "field house" sites are defined as sites with no evidence of a pit house but with some surface debris indicative of one or two rooms. Most are quite small and could at best house a single household during some portion of the year (Kohler 1992, p. 626).

It is admitted that for the DAP "...no actual field locations, other than those presumably adjacent to field houses, can be recognized in the local archaeological record" (Kohler 1992, p. 626). Kohler (1992) talks about field houses but provides no archaeological definition of a "field" independent of the identification of "field houses." He creates a simulation showing that the mean distance from field houses to villages of zero to 1.5 km, but does not address the distance from field houses to fields (since he identified no fields archaeologically). Fields were defined apparently as those areas of suitable cold-air drainage, soils quality and climate.

Lang (1995) describes archaeological sites in an area of about 13 ha immediately to the southeast of the northeast corner of the San Marcos Pueblo of New Mexico. He identified 15 different types of features, such as ak-chin fields, linear border fields, and gravel mulched fields. The term "ak-chin" is apparently a Papago Indian word meaning "arroyo mouth" (Bryan 1929, p. 49), an agricultural setting where water confined to an alluvial channel in southern Arizona debouches on to a flatter floodplain and spreads out. Also identified are three "field structures" (Lang 1995, p. 58) "small rectangular buildings formed of one to three rooms, having cobble foundations or combination foundation-stem walls." Artifacts associated with these include pottery, lithic debitage, a mano, and heat treated rocks, suggesting a range of activities (Lang 1995, p. 58). He describes their purpose as "temporary storage of crops, water and tools, food preparation and shelter when needed." There are literally dozens of small agriculture plots or fields in the area but only three field structures. It's not clear how far these field structures are from the closest clear agricultural feature (like a gravel mulched plot or linear border) plot—but certainly since there are so few of the field structures most of the agricultural fields are likely less than ~300 m away from the nearest field structure.

Archaeological survey was conducted on a prominent mesa top about one kilometer from the La Bajada Ruin, which is a ~500 room pueblo dating A.D. 1325–1650 in New Mexico. Extensive "dry farming fields" were found, consisting of a discontinuous distribution of cobble pile arrays and grid patterns over 2000 ha of the mesa (Herhahn 1995, p. 78). These cobble pile arrays and grid agricultural features are often associated with "field structures" which Herhahn says are similar to the windbreaks Mindeleff (1891) described for the Hopi. "Generally one or two field houses are associated with each discrete cobble pile/grid area, constructed of basalt cobbles and boulders, generally two or three courses high" (Herhahn 1995, p. 78). Some are crescent shaped and open to the northwest to protect from the prevailing wind; there are also one and two room structures in some areas. Surface artifact are found near the structures including ceramics, lithics and ground stone. Some of the (one or two room structures?) may have been habitations; the open-ended crescent shaped ones may have been rudimentary shelters used while tending crops, scaring predators, and maintaining features. In any case, they seem to have been located quite close to unambiguous agricultural fields or plots at least; perhaps within 100 m.

Contemporary Anglo dryland farmers also place field houses near fields. Linda Honeycutt (1995) presents a description of modern Anglo dryland gardening techniques as currently practiced in southwest Colorado. This part of Colorado receives about 357 mm of precipitation annually and ~120 frost free days per year. Honeycutt states that the precipitation arriving as snow is critical to maize since it gives the soil enough spring moisture to germinate maize. Honeycutt argues (1995, p. 372) that ancient maize farmers likely stayed nearby to tend the crops, eliminate weeds, and deter predators. "When populations were small, a single habitation could easily be situated next to a garden plot. As population increased in these hamlets or villages, the number of fields needed to sustain people increased, and correspondingly these would have been at greater and greater distances from the residential village. "Field houses would have been useful as guard stations related to animal predation, places for tool and equipment storage, and rest areas." As corn ripening approached, field houses were probably continuously occupied, in response to the voracious appetites of deer, crows, rabbits, raccoons, porcupines and other animals for ripe corn (Honeycutt 1995, p. 372).

3.2.7 Late Holocene Climate in the Study Area

The period of Pueblo I and II in the American Southwest generally and Mt. Trumbull in particular was generally coincident with the "Medieval Warming" period of Europe A.D. 900–1300 (Cook et al. 2004, Woodhouse et al. 2010). It is generally believed that a very severe widespread drought occurred across the ancient Southwest beginning perhaps in the late 13th century (Woodhouse et al. 2010). This drought is sometimes referred to as the "Great Drought" and is thought to have lasted 35–50 years. Other severe droughts although of a shorter 10–20 years duration occurred before, interspersed with periods of greater precipitation. Most if not all of these extended periods of drought coincided with warmer temperature (Woodhouse et al. 2010) although not as warm as the last 30 years.

These drought periods would likely have made it imperative that Pueblo I and II farmers took full advantage of small variations in temperature and soil moisture to survive (Cordell 2009, p. 383–385). Long periods of inadequate precipitation would have likely been more damaging than warmer temperatures, since increased temperature may have resulted in longer growing season in this area. Unpredictability or lack of sufficient precipitation at critical parts of the growing season might have forced even greater dispersal of settlements in a given region, requiring inhabitants to travel greater distances from fields to habitation, or to increase the number of farm plots in any given area.

3.3 Methods

3.3.1 Study Sites

A portion of the study area was identified for intensive study for development of the predictive model (labeled as area A on Fig. 3.3.) Prior archaeological survey in the 1970s (Moffit and Chang 1978) and more recent survey and testing as part of an on-going archaeological field school (Buck 2002; Buck 2005, 2006, 2010) has shown a number of larger C-shaped pueblos in this area, and intensive survey was started in 2006 to record all significant sites. The summer field school systematically surveyed these areas at very close spacing to identify virtually all sites, including the more ephemeral artifact scatters. The area was selected in part because thorough archaeological survey had already been completed. The models were constructed using data from a "training" area ("C" of Fig. 3.3, sections 10, 11, and part of 12 T. 34 N, R. 8 W), then tested against a different portion of the study area where no archaeological survey had yet been done, i.e., Area "B" (sec. 15 T. 35 N, R 8 W) of Fig. 3.3.

The end product of the image data and ground based investigations are predictive model(s) specifying optimum locations for maize production within the study area. These are not strictly "run-off" models, since infiltration, surface roughness, soil stratigraphy are also included, but are more properly described as

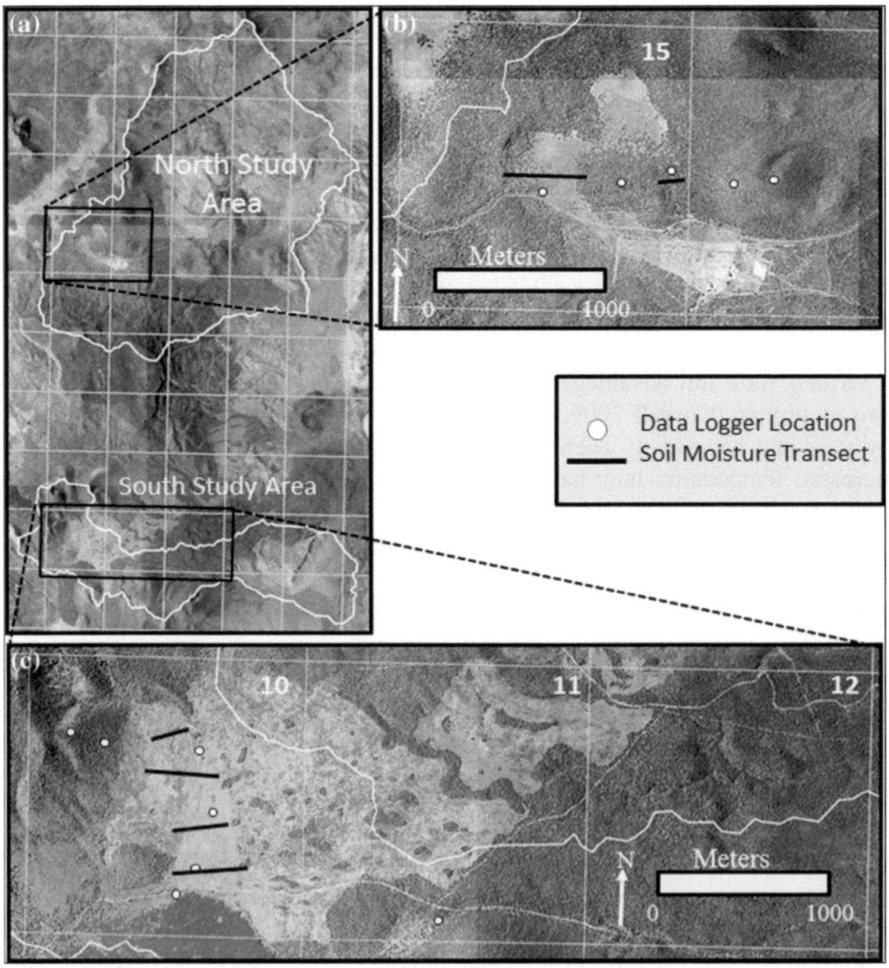

Fig. 3.3 Northern and southern study areas (a) showing locations of temperature data loggers and soil moisture transects, (b) shows the test area and (c) the "training" area

"soil moisture" models or "water allocation models" (Rhode 1995). The models are described below as the "restrictive" model and the "fuzzy logic" model. Both models use the same attributes (or layers) of surface gradient (topography), soil type, temperature, solar illumination, and estimates of soil infiltration and runoff (derived from a separate hydrological model called FLO-2D, described below).

We use remotely sensed data of a variety of types in conjunction with ground based observations to identify optimal areas for prehistoric agriculture using a grid of about 30 m × 30 m across the study area. We then compared the distribution of sites (especially so-called field houses), with the locations of optimal field areas.

Below we describe how GIS layers were created from the attributes of interest, and these layers turned into a several different "optimality" models. Based on literature of maize ecology and traditional maize agricultural practices from ethnographic literature (especially from the Hopi), the favorability or optimality of portions of the study area based on each of the characteristics was ranked generally from best to worst with a rating of 3–5 levels. These spatial layers were then overlain in a 30 m grid cell size (pixel) and maps generated showing the spatial distribution of most optimal to least optimal areas. This modeling effort was limited to two small drainages on the study area for which hydrological models were generated (building the hydrological models takes significant computation and staff time). We then examined the distribution of recorded archaeological sites of various types in relation to field optimality zones. We calculated how many of each site type is found. Site types include: artifact scatters without architecture; structural sites with one or two rooms; multi room sites with three to eight rooms and light artifact scatter; and larger C or L shaped structural sites with more than eight rooms and abundant artifact scatter. We then evaluated how many of these sites are found within 100 and 200 m of the potential maize fields. Finally using the southern study area as a data set we predict the distribution of sites for a previously unknown area given the specific optimality zones modeled for this small area.

The attributes are generally grouped into two major types: those related to temperature and those related to water. Water surface flow and water infiltration, and soil moisture, and soil type are on the "water" axis; surface gradient, solar illumination, and radiant temperature (from ASTER, the Advanced Spacebourne Thermal Emission and Reflection Radiometer) are on the "temperature" axis.

3.3.2 *Rationale for Choice of Analytical Unit (pixel) Size: How Big were Field Areas?*

A key theoretical and practical decision in our analysis is the selection of the appropriate scale/size of the analytical unit and the identification of an appropriate defensible distance measure between fields and structures (like field houses). How big were Anasazi fields, and what distance seems reasonable for field houses to have been located from these? Archaeological and ethnographic evidence described below suggests appropriate sizes for both these.

Traditional Hopi horticulturalists cultivated areas ranging from ~6–100 ha depending on the size of the watershed feeding the maize fields (Hack 1942). Bryan (1929) shows field size from areas in Zuni Pueblo and others areas before irrigation. He indicates that about 67 ha combined for Zuni Valley (in three plots, the largest perhaps 40 ha, the smallest perhaps 12 ha) although "only parts of these areas were actually planted and still smaller parts were harvested" (Bryan 1929, p. 450). A nearby Navajo corn field had a size of 22 ha. Archaeological rock gardens along the lower Rio Chama Valley of New Mexico (Anschuetz and Maxwell 1986) range in size from the very smallest (several square meters) to the largest single identifiable field is about 30 m × 40 m. Aggregates of gardens in this area can extend for 600 m along the Rio Chama. Kohler (1992) uses a size of 4 ha for his agricultural simulation.

Our analyses considered a minimum size of maize field for analytical purposes to have been about ~30 m × 30 m; 0.22 acres or 0.09 hectare. This is reasonable given the ethnographically known sizes of Southwest Pueblo maize fields, and also suitable for remote sensing analyses. Imaging systems such as MASTER (MODIS/ASTER Airborne Simulator, the aircraft mounted test-bed version of ASTER) has a pixel size of 5–50 m; ASTER data ranges from 15–90 m; Landsat data is 30 m, the hydrological modeling data is about 30 m pixel size, and DEM data (digital elevation model) is 10 m resolution. For most remote sensing data this results in a sufficient number of pixels (except for ASTER) for statistical analyses. The Flow 2D hydro model uses ~30 m pixels.

We use a statistical test called Chi square or contingency table analysis to evaluate the co-occurrence of optimal zones with different types of archaeological sites. The Chi square test is a very general test to evaluate whether or not frequencies which have been empirically obtained differ significantly from those that would be expected under a certain set of theoretical assumptions (Blalock 1979). Chi square is well suited to the problem here; i.e., are certain site types disproportionately found near theoretically defined "optimal" maize fields? To conduct a Chi square analyses the data was collapsed into 2 × 2 contingency tables (in part due to low counts or in one case a zero count) in the cells. We generally combined one and two room sites with scatters, arguing that scatters could be the only visible remnants of a perishable field house which is now gone. So-called "field houses" in ethnographic literature are known to be constructed of perishable materials (Mindeleff 1891 shows some at least). After 1,000 years such a field house might look like a small lithic/ceramic scatter. Multi-room sites were combined with large sites, since many of these multi room sites had up to eight rooms and/or had a surface depression indicative of a rather substantial occupation.

3.3.3 Model Data Layers

The product of the image data and ground based investigations is a model specifying optimum locations for maize production within the study area. This is not strictly a "run-off" model, since infiltration, surface roughness, soil stratigraphy etc.

will also be part of the model. It is more properly a "soil moisture" model or a "water allocation model" (Rhode 1995). We used FLO-2D to model surface water hydrology in two basins in the study area. The model is constructed using data from a relatively small intensively studied area shown as area "C" in Fig. 3.3 called the "Training" area. Archaeological surveys are completed and we know where all the sites are. All available remote sensing data are analyzed; DEM's are examined and incorporated into the run-off model. The model is then tested against a currently unknown area at the northern boundary of the study area labeled B (Fig. 3.3). We examine the relationship spatially and statistically between archaeological sites and modeled optimal maize fields from area C, then examine whether or not the relationship holds true in a previously unknown area B.

We collected and used ground-based field measurements and satellite image data to produce model input "layers" that were then input into a GIS where we could look for patterns in ancient use, surface characteristics, and composition to predict possible ancient maize agricultural fields. One "training" area was initially used to develop the models; this is "area C" of Fig. 3.3 (sections 10, 11 and part of 12, T. 34 N, R. 8 W). Input data layers for the model included: (1) early spring maximum solar illumination, (2) surface gradient, (3) surface radiant temperature, (4) water surface flow collection, (5) water infiltration, and (6) soil suitability for maize production. Once the models were developed using data from the "training" area, the models were then analyzed with reference to the archaeological field data to determine the proximity of different archaeological sites to probable agricultural sites. We used the Chi square statistical test to evaluate if different sites types were found preferentially close to optimal zones. We then applied the same models to a previously unknown area (area B in Fig. 3.3) and examined whether or not similar kinds of archaeological sites were associated with the same types of optimal areas modeled in this new area.

We note here that two of our six model data layers are directly tied to water availability, which for all practical purposes makes this factor twice as important as any other factor. The importance of water in the successful germination and harvesting of sufficient yields of maize has been noted ethnographically (Hack 1942; Bryan 1929; Bradfield 1971; Dominguez and Kolm 2005) and also in other predictive models of agricultural potential (such as Dorshow 2012). When modeling the agricultural potential of Chaco Canyon during the Bonito Phase, Dorshow (2012), p. 2102 uses six criteria of varying weights or importance. Three of these are related to water for a total weight of about 38 % of the model. It is true that many traditional agriculturalists such as Hopi and Zuni take a variety of factors into account when selecting places to plant (Dominguez and Kolm 2005) including soil characteristics, slope, and other factors, it is often water availability that is limiting in the production of maize. Dominguez and Kolm (2005) are careful also to indicate that their model is not simply a re-wording of traditional Hopi practices and is not unique to the Hopi, but has wider application because it examines the underlying abiotic and biotic variables relevant to maize agriculture in the Southwest. Similarly, Dorshow (2012) develope predictive model independent of the distribution of archaeological sites in his study region, identifying abiotic factors that may have combined as best places for maize agriculture without relying on locations of known sites to assert "fields."

Water availability is accessed using current climatic information in combinations with surface run-off models. ASTER satellite data with 90 m resolution is used to build a model of surface radiant temperature using thermal infrared energy captured in five wavelengths: 8.3, 8.65, 9.1, 10.6, and 11.3 μm. A DEM is used to construct a solar illumination or insolation layer. Soil characteristics important for ancient agriculture derived from the US Natural Resources Conservation Service using attributes such as organic matter, available water capacity and soil profile texture characteristics are incorporated into the model. The Hopi Agricultural Project (Dominguez and Kolm 2005) found that stratified soil profiles with alternating course and fine layers were preferentially selected for maize agriculture because of their better moisture retaining properties. Also historical Landsat and ASTER data are used to identify areas where snow accumulations typically occur in the area. Snow melting rate may be roughly estimated by looking at images immediately after snow-fall rates since snow depth cannot be directly ascertained.

Early Spring Maximum Solar Illumination. The early spring maximum solar illumination was determined by sequentially artificially illuminating a 10 m DEM for mid-day during the month of May. Late April—early May are when the last freeze typically occurs in the study area and is when maize planting would most likely occur. To simplify the model calculations data was broken into 6 classes ranging from white (areas that received the most sunlight) to black (those that receive the least).

Surface Gradient. Surface gradient was calculated using the 10 m DEM and reflects the propensity for agricultural fields to be located in areas with lower gradients, i.e., minimal or no slope.

Surface Radiant Temperature. Surface radiant temperature was determined using ASTER thermal data (AST08 data product). The data were collected 13 April 2001 and represents typical Spring day-time temperature radiant temperature distribution in the study area. We used the Nixon Flats and Mt. Logan weather station data to adjust the ASTER radiant temperatures to more accurately reflect air temperature. ASTER temperatures from different dates at these two stations (17 for Nixon Flats and 13 for Mt. Logan) were plotted against air temperature measured at the weather stations (Fig. 3.4). The ASTER temperature data, which tends to be warmer than the measured air temperature, were adjusted using these regressions. Warmer areas are considered better for agriculture than colder areas.

Hydrological model (Meyer et al. 2010). The hydrological model FLO-2D (O'Brien et al. 2007) is used to evaluate the hydrology of two watersheds in the Mt. Trumbull area of northwest area of Arizona (Fig. 3.3b, c). Both topography and soil conditions are considered in the hydrologic model for this study. Data used for the precipitation is taken from both the NOAA atlas and the Natural Resources Conservation Service (NRCS). A discritzed two dimensional model that considers elevation data and the spatial distribution of the soil parameters at a ~30 m square grid is used. Data used for both the elevation and soils information is taken from readily available sources on the internet. The soils information

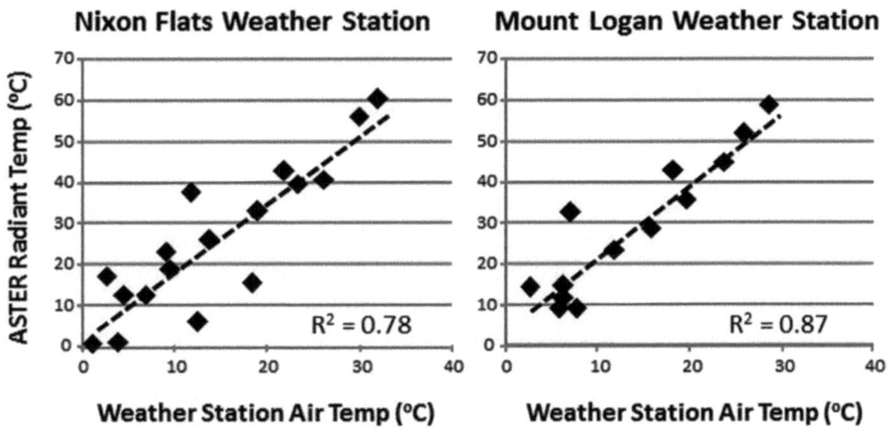

Fig. 3.4 Plot of ASTER radiant temperature versus meteorological air temperature for Nixon Flats and Mount Logan meteorological stations

is further developed using standard soil science methods and spatially distributed using ESRI's ArcMap (version 9.3.1, by Environmental Systems Research Institute, Inc., Redlands, California).

Data used for the final models considers both rainfall and soil variation. Variation in the rainfall intensity is developed through the use of an NRCS s-curve and point precipitation number. Methods used for developing and varying the soil infiltration parameters are similar to those presented in Meyer et al. (2010). Development of the rainfall input to the final models is similar to the method prescribed by the Clark County Regional Flood Control District's Drainage Design Manual (1999). Data used in the FLO-2D model was taken from several sources and processed with ArcMap. Topographic information used in the hydrologic model was taken from the United States Geologic Survey's (USGS) National Map Server at http://seamless.usgs.gov/website/seamless/viewer.htm. The raster information is downloaded in the National Elevation Dataset (NED) format, and is the best seamless dataset available for the conterminous United States. Additionally, the spatial distribution of the soil surfaces in the modeled watersheds is taken from the United States Department of Agriculture's (USDA) NRCS (formerly the Soil Conservation Corps) Web Soil Survey at http://websoilsurvey.nrcs.usda.gov/app/WebSoilSurvey.aspx. The NRCS website provides soil data in the ArcMap shape file format that spatially locates the soil surfaces, and in a text document that provides the soil properties of each soil type.

Spatial variability of soil properties in the watershed is defined by the USDA NRCS shape file. The watershed has six well defined soil surfaces, each of which were parameterized using the G-A (1911) soil parameters developed from the NRCS soil textural data presented in the WebSoil survey. The G-A method was chosen as it links soil properties developed with a pedo-transfer function (PTF) to variables used in the final runoff model. The method relates the saturated

Hydraulic conductivity (K_s) to rainfall intensity. The infiltration rate (*f(t)*)is described by the G-A equation (Mein and Larson 1973) as:

$$f(t) = K_s \left[1 + \frac{\psi_s(\Delta\Theta)}{F(t)} \right]$$

where, K_s is the saturated hydraulic conductivity, ψ is the soil suction at the wetting front, $\Delta\Theta$ is the soil water deficit and *F(t)* is the cumulative infiltration. The G-A parameters used in the FLO-2D model are developed through a multi-model platform similar to that presented in Meyer et al. (2010).

The first step is to evaluate the soil data from the NRCS soils report using a PTF. The PTF is an approach to estimate soil hydraulic properties from relatively simple-to-obtain field measurements, like soil texture (Arya and Paris 1981; Arya et al. 1999). The concept has been used for a variety of field scenarios dealing with infiltration and water movement (Rawls et al. 1989; Schaap et al. 2001; Parasuraman et al. 2006), and for parameterizing numerical models using different approaches (van Genuchten 1980; Mualem 1976). The PTF used for this study is the software package Rosetta (Schaap et al. 2001) which provides a well established method to determine the van Genucthen soil parameters used in the next step of the process that develops the wetting front soil suction (ψ) and the soil water deficit ($\Delta\Theta$).

The van Genuchten (1980) parameters developed with the PTF were used to evaluate the $\Delta\Theta$ and ψ for each soil type. The van Genuchten parameters of residual water content (θ_r), saturated water content (θ_s), alpha curve shape parameter (α), n curve shape parameter (n), and the tortuosity parameter (l) are all input variables used to calculate $\Delta\Theta$ using equations developed by Mualem (1976) and van Genuchten (1980). The ψ is then calculated using the equation developed by Neuman (1976). The values $\Delta\Theta$ and ψ assumed for this study are at field capacity of 100 kPa or 100 cm of matric potential.

The precipitation values used in the model are simplified from an actual rainstorm. The model assumes consistent rainfall across the entire watershed. Although the precipitation does vary with time in the model, at any given point in time the rain is consistent across the entire watershed area. The storm considered for this study is an NRCS 24-h Type II Storm with a recurrence frequency of two years. The point precipitation value referenced from the NOAA website is 39 mm (NOAA, 2011). This value indicates that for a two year frequency storm with a 24-h duration a total depth of 40 mm of rain will fall on the study watersheds. The 24-h storm was selected as this more closely resembles the type of storm that is experienced in Northern Arizona during the winter months, and is the design storm recommended by the NRCS for this area. The precipitation event is sufficient to produce runoff in both modeled watersheds.

Soil suitability for maize production. Soil suitability for maize production was derived from the NRCS soil survey using attributes such as organic matter, available water capacity and soil profile texture characteristics. The Hopi Agricultural Project (Dominguez and Kolm 2005) found that stratified soil profiles with alternating course and fine layers were preferentially selected for maize agriculture because of their better moisture retaining properties. Using these criteria, soils were classified as best, moderate (sub-optimal but capable of supporting agriculture), and poor. Soils data were taken directly from the NRCS web soil survey data. The characteristics of soils and substrates have a considerable influence over the suitability to grow prehistoric maize (Dominguez and Kolm 2005). These include porosity, volumetric soil moisture, hydraulic head, alternating layers of sand and silt, and others. To estimate the most suitable soils for maize agriculture we reviewed literature about maize ecology and physiology (i.e., Toll 1995 and Dominguez and Kolm 2005) and then ranked the suitability of soils in the Mt. Trumbull area.

The web soil survey of the Natural Resources Conservation Service soil survey, http://websoilsurvey.nrcs.usda.gov/app/HomePage.htm) was used to identify and classify soils in the region. The NRCS defines soil units based on aerial photographs with one or at most few soil pits excavated in representative areas. Available on the website are a number of important characteristics of soils, including soil physical properties, soil chemical properties, suitability for agriculture and others. Some of these attributes are the same or similar to those in literature about traditional Hopi and Zuni agricultural practices. For example, studies of productive Hopi maize fields show preference for stratified soils—soils with alternating layers of sand, silt and clay (Dominguez and Kolm 2005). Two other attributes listed in the soil survey for Mt. Trumbull also mentioned by Domniguez and Kolm include available water capacity and percent organic matter.

3.3.4 Data to Validate Model Inputs

Additional measurements were made to allow validation of the above inputs. These include field measured air temperature and soil moisture, remote measured snow pack and Spring green-up, and ancillary climatic data from long-established weather stations.

Temperature data loggers were maintained in the field for 12 months to: (1) validate the temperature variation seen in the ASTER image data, and (2) determine the number of frost-free days at different elevations in both study areas. Five loggers were located in and near Section 15 in the northern area while seven loggers were deployed in the south (Fig. 3.3). Temperature measurements were made hourly. An additional logger was co-located with the Nixon Flats weather station to validate the accuracy of the data loggers measurements (Fig. 3.5).

Soil moisture transects were measured in the field using a FieldScout TDR 100 with both 7.6 and 20.3 cm probes and measurements made every 24 m along the transects in the Southern Area (every 50 m in the Northern transects). Ideally, we would have soil moisture loggers imbedded ~0.5 m depth. However, because of the sensitive archaeological importance of the area, we were not permitted to perform any excavations. These shallow probe measurements were the best we could do under the circumstances. These probes did not allow us to measure within the

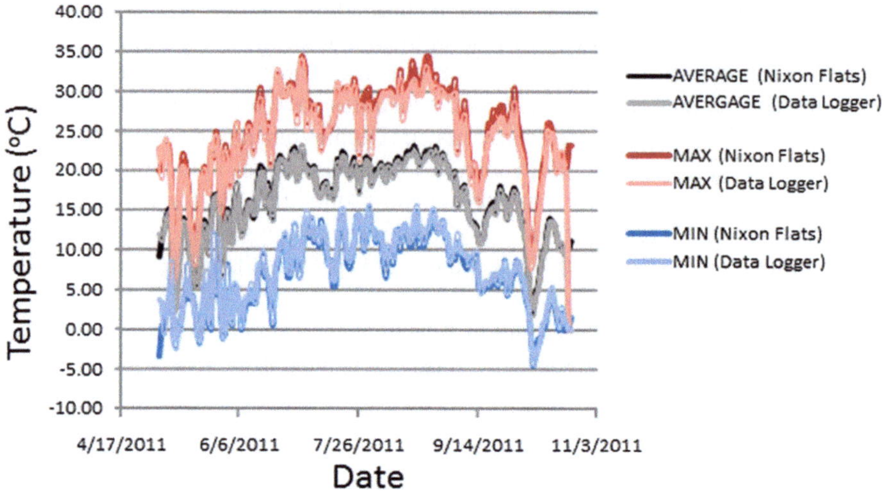

Fig. 3.5 Comparison of data logger temperature with measured air temperature from Nixon Flats meteorological station

stratified layers at depth thought to be important in soil moisture retention and, therefore, the data should be considered with that in mind. Field measurements were made in May and August of 2010 (Fig. 3.6).

Historical Landsat and ASTER were used to identify areas where snow accumulations typically occur in the area (although no layer was created from these data for input to the models). Snow melting rate may be roughly estimated by looking at images immediately after snow-fall. These data may only be useful to estimate these rates as snow depth cannot be directly ascertained.

Areas of strong Spring green-up were identified using two ASTER scenes, one late fall and the other early spring. The scenes, taken on 13 April 2001 and 23 November 2001 were processed using a simple NDVI (Huete and Jackson 1987), registered and compared. Although useful, the lack of temporal resolution during the Spring limits this approach as we could easily miss a green-up that occurred after the image was taken. However, the approach was useful early in the study to identify potential study areas.

3.3.5 Archaeological Survey

A long term program of archeological investigations was begun in the Mt Trumbull area in 2001, primarily using an undergraduate class in field methods in archaeology from Nevada State College. The field school instructed students in archaeological survey. A large (relatively) Puebloan architectural site known as "Ken's Big E" (it was named after a volunteer and is in the shape of a capital

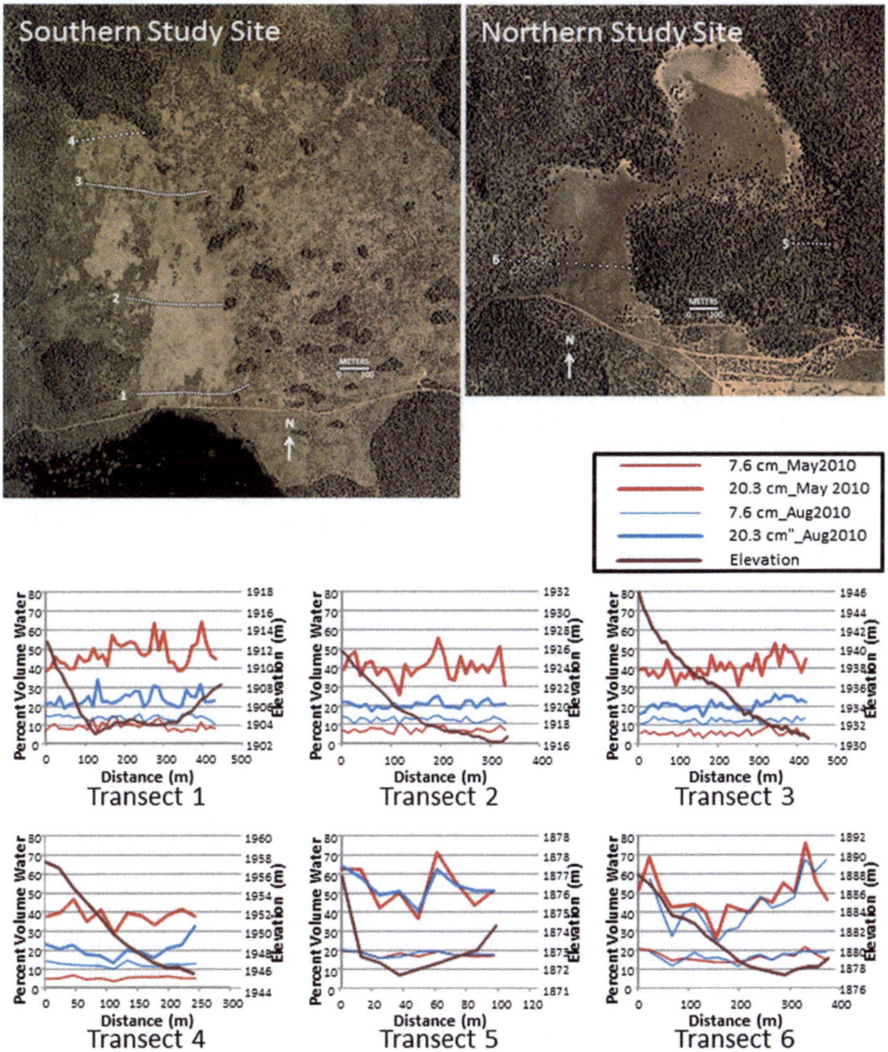

Fig. 3.6 Soil moisture transect data from field scout TDR 100 soil probe. The figure shows soil moisture at depths of 7.6 and 20.3 cm in May and August

letter "E") was found in the first year of the field school and recorded in detail in Section 11; we then decided to complete comprehensive detailed survey in this entire section during the period 2003–2007. This included an area known in prior ecological reports as the Bird Plot survey area, where in the 1970s ornithologists had tagged and labeled trees in a grid pattern and recorded nesting trees of birds. Within a few weeks after survey began it was clear this area contained an unusually high number of previously unrecorded archaeological sites and it was determined to survey the entire section. In 2008–2011 the adjacent section 10 to the west was also surveyed for essentially the same reasons. A portion of the western part of section 12 was surveyed in 2011.

The areas surveyed do not represent a randomly selected portion of the study area. It is likely that this area contains an unusually high number of archaeological sites. The only archaeological survey in the Mt. Trumbull area that perhaps can be considered random is that conducted in the 1970s by Moffit and Chang (1978), although even this is perhaps strictly not random (see Altschul and Fairly 1989).

The "test" or validation area of the predictive model was chosen as section 15 T. 35 N R 8 W north of Mt. Trumbull (area B on Fig. 3.3). The purpose of this (also non randomly chosen) area was to provide an initial test of the predictive model (the test is described below). This area was selected because it was not previously surveyed, was located on the opposite side of Mt. Trumbull, and was accessible to survey crews. Results of that survey will be discussed below.

Methods of survey within selected areas. The Bureau of Land Management usually classifies archaeological survey as Class III—the most intensive coverage—if it uses pedestrian transects spaced ~30 m or less apart. As part of the field school we conducted such survey with a line spacing of 5–10 m, with the goal of finding all significant sites in the survey area. Any architectural site or site with recognizable features was recorded as a discrete site; each scatter of artifacts was given a site number when more than 30 items were found within a 15 m radius of each other. When an artifact was located the survey crew stopped and marked their lines; a thorough search of the immediate area was conducted looking for artifacts. When in doubt a 30 m tape was used to verify if more than 30 artifacts were present in a 30 m diameter circle and the site was formally recognized with a site number. Significant isolated artifacts were also recorded (such as obsidian flakes/points [for geochemical sourcing], unusual artifacts like whole vessels or rare objects like stone adzes).

A State of Arizona "site card" was completed for each site found. The "site card" is a State of Arizona mandated site recording form to be completed for each archaeological site found or re-recorded. It includes a sketch map to scale, a description of setting of the site, and a rough count of artifacts seen on the surface. If any structures or features were present, a separate sketch was made of that. Different more intensive procedures were followed to collect representative artifacts from selected sites (not discussed here, but see Buck 2005, 2006).

Chronology. Importantly, chronology is very poorly constrained. For the purposes of this paper, all sites are essentially contemporaneous (Publeo II). From surface artifacts in well-studied regions of the Southwestern US, it is often possible and routinely done to develop a chronology based on pottery styles and projectiles to some extent (for example Cordell 2009; Lang 1995; Lyneis 1995). Most of the pottery in the study area seems to be either late Pueblo I or Pueblo II, dominated by Moapa gray wares and common black on white sherds most with Pueblo II design styles (Buck and Sakai 2008). Some sites have a few black-on-red sherds or Tsegi orangeware. At this time, due in part to lack of controlled excavation or thermo luminescent (TL) dating of sherds, it is not possible to control chronology better than ~200–400 years of the actual date. So even though it is likely that occupation in this immediate area spans at least several hundred years, all the sites are lumped as one time period. We recognize that not all the large habitations or the smaller "field houses" were likely to be occupied simultaneously. The converse is much likelier. The lack of detailed chronology is likely to obscure much important change over this period. It is possible that even the largest C shaped pueblos are as ephemeral as the smallest field houses. In other words, each place may have been occupied for a relatively brief time, perhaps 10–30 years, with new C-shaped pueblos being built after the earlier one was abandoned or perhaps simply to be closer to important resources like fields or water.

3.4 Results

3.4.1 Model Attributes or Layers (Predictive Model Inputs)

Early spring maximum solar illumination. Solar illumination is shown in Figs. 3.7a, 3.8a for about a month in late spring for a single year. If this were run continuously over a year and summed, it would show the total solar energy or insolation at any given spot; higher total values would indicate more heat energy there than in other spots.

Surface gradient (topography). Generally speaking, lower elevation areas are warmer than higher elevation area (Figs. 3.7b, 3.8b). This is clearly true when comparing Mt. Trumbull schoolhouse and Tuweep meteorological stations with Mt. Logan and Nixon Flats. However, it is likely that some small valley bottom areas are colder than might be expected because they are cold air sinks. Colder air from higher elevations moves down slope and settles in enclosed valleys sometimes causing a temperature inversion. This is seen somewhat in our temperature data loggers and is obvious when looking at data from Mt. Logan and Nixon Flats weather stations, where temperatures are actually colder at Nixon Flats even though it's 213 m lower than Mt. Logan. Figure 3.4 shows a comparison of temperature from the meteorological towers for a selected period versus temperature recorded from data loggers for the same period. At certain times of day lower elevation basins are colder than areas slightly higher in elevation apparently because colder air moves down slope and settles in topographic "lows." This "cold air sink" effect is shown in Fig. 3.9 where the number of frost free days is plotted against elevation in Rabbitbrush Valley in the "training" area (Fig. 3.3c) and section 15 (the test area Fig. 3.3b). In both cases the colder temperature (and hence the shortest growing season) is found on the bottom of the small valleys, while somewhat warmer temperatures tend to be found on gentle slopes above the valley floor. This could potentially shorten the growing season in valley bottoms by a number of days, putting a maize crop at risk.

Despite this fact, we do not have sufficient temperature information to distinguish these potential cold air traps from other areas. We assume in the model that lower elevation areas are more favorable than higher elevation areas. However, it is likely that those areas on gentle slopes above the valley bottoms might be more suitable than slightly lower lying areas which act as cold air traps.

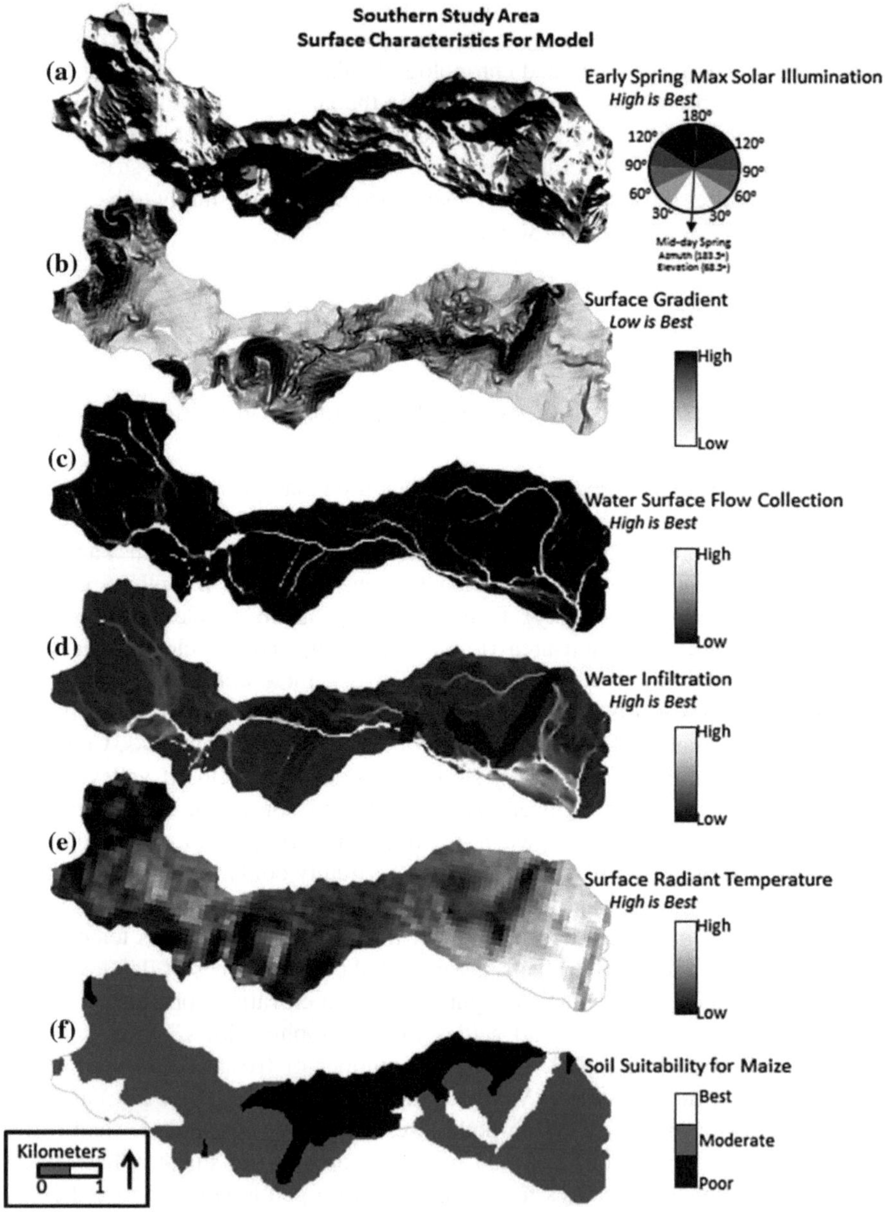

Fig. 3.7 Surface characteristics for modeled southern study area

Surface radiant temperature. Surface radiant temperature is shown in Figs. 3.7e and 3.8e. As mentioned above these are derived from the ASTER scene of April 13 2001. This should be considered a relative temperature image showing those areas in

Fig. 3.8 Surface characteristics for model northern study area

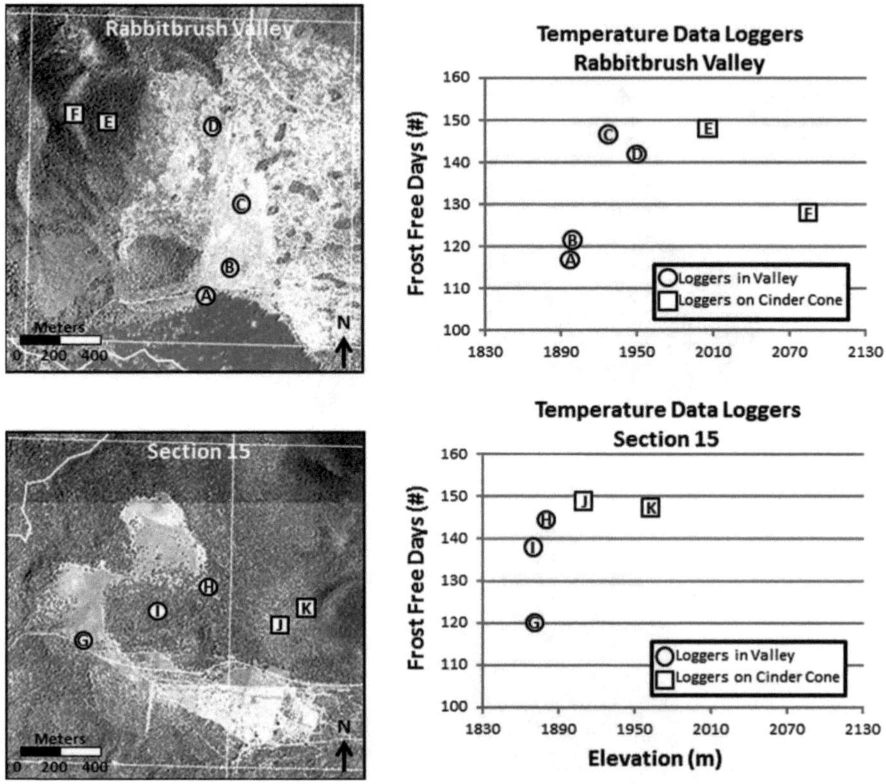

Fig. 3.9 Data from temperature loggers in a part of the southern study area (Rabitt brush Valley) and in the northern study area (Section 15) showing relationship of elevation to number of frost free days

white (brightest) as warmest and those in black as coolest. It is believed that this same pattern will hold true for other times of year and that the pattern holds for even for the time of day when temperatures are coldest ~4 AM local time. That is, areas shown as coldest in the scene collected 11 AM will also be coldest at 4 AM. Figure 3.5 relates measured air temperature from met stations with ASTER radiant temperature.

Water surface flow and water infiltration. The study areas include areas with high potential for soil moisture. Figures for two separate views of the watershed response for each study area has been included (Figs. 3.7c, d and 3.8c, d). The first shows runoff in terms of flow depth above the soil surface. Values for flow depth shown on the figures assume both infiltration excess at each of the grid cells as well as run-on from upstream grid cells. The figure shows areas where flows tend to converge; in-other-words where streams are forming due to the runoff generated in the model.

The second set of figures (Figs. 3.7d and 3.8d) considers the amount of precipitation infiltrated to the soil in terms of depth into the soil (below the soil surface). The values shown in the figures also show areas were infiltration is occurring due to rainfall at the specific cell as well as run-on from upstream grid cells in the model. The infiltration and runoff areas are a function of the topography and soil texture (proportion of silt + clay).

There are areas in both the south and north watersheds where topographic and soil conditions may result in greater soil moisture (Figs. 3.7c, d and 3.8c, d). Both watersheds show areas where flows spread out beyond the 30 m grid cell. The soil has the ability to absorb water in these areas also. High flow depth and high infiltration areas are present in the loam areas.

Soil suitability. All the soils in the two drainage basins where the FLO-2D hydrological models was applied were examined and placed into one of three categories: best, intermediate, and poor. Best soils were those with three or more alternating layers of sands, silt, and clay; the highest available water capacity; and the highest organic matter. For example Showlow soils with 1–15 % slopes have four strata some with 43 % sand and others with 6 %, an available water capacity of up to 53 mm of water per inch of soil; and up to 3 % organic matter. "Poor" soils include bare lava flows such as the Little Springs flow; those with the least ability to hold water; and those with little organic matter.

Only one of the soil units in the study area is indicated by the NRCS as at all suitable for growing corn–Unit 62 Sponiker gravelly loam, 15–40 % slopes has an estimated yield under optimal conditions of 9.58 m^3/ha. According to NRCS this value is "…an estimated average yield per acre that can be expected of selected irrigated crops under a high level of management."

3.4.2 Results of Archaeological Survey

Archaeological survey has been completed in sections 10, 11, and the western ¼ of section 12 T. 34 N, R. 8 W (Fig. 3.3c). A total of 142 sites are now known from this area (Fig. 3.10, Table 3.1). This includes 50 sites termed one or two room structures ("field houses"), 46 multi-room structure, 13 larger C or L-shaped pueblos, 29 scatters, 2 rock art sites and 2 historic sites. Due to federal and state laws we are prohibited from accurately depicting the locations of archaeological sites in the study area (to prevent possible looting).

Site types were defined using field notes and the site coding forms based on survey information collected in the field. Structures were identified by locating any foundation stones visible during ground inspections—no subsurface testing or scraping was done, and little vegetation was removed. Site sketch maps were made after identifying structures and site boundaries.

A "field house" was recognized if there were one or two distinguishable rooms either connected or isolated, whether these were circular or rectangular (Fig. 3.11). The structures were typically built of loose basalt rocks and cobbles sometimes several courses high. Often a light surface scatter of artifacts (pottery sherds and chipped stone) was also co-located with the structure(s). Any site with an associated surface depression, likely a pit house or possible ceremonial structure—was

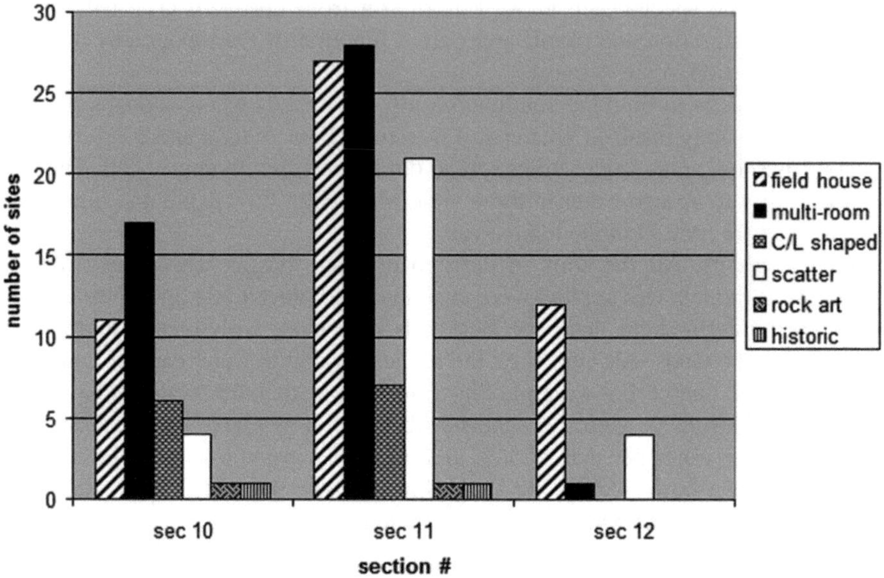

Fig. 3.10 Bar chart showing counts of different types of sites in the 'training' area C

Table 3.1 This table shows all sites found in sections 10, 11, and 12, T. 34 N R8 W (Fig. 3.2c). Some of these are outside the modeled hydrological model or basin used in the analyses

	Field house	multi-room	C/L shaped	Scatter	Rock art	Historic
Sec 10	11	17	6	4	1	1
Sec 11	27	28	7	21	1	1
Sec 12	12	1		4		

not included with the "field house" category since this implies a more substantial, longer term habitation. Such sites were included in the "multi-room category" even if there were only two rooms (plus the surface depression). Multi-room sites are those with more than two clear rooms or structures up to perhaps ten rooms with or without surface depressions and with substantial scatters of a wide variety of artifacts such as pottery, lithics, and ground stone.

The largest habitation sites recognized in Mt. Trumbull are C or L (or in one case E- shaped) pueblos of 10–20 rooms often arranged in a clear C or oval shape and with dense concentrations of pottery and other artifact types on the surface. An example is shown below (Fig. 3.12); this is site AZ:A:12:136 (ASM) also known as "Ken's Big E" pueblo. Scatters are those sites with no structure visible—just a more or less dense scatter of pottery, lithics, sandstone fragments and grinding stones. These have a density of at least 30 objects in a 30 m diameter circle. Two rock art sites were found—each consisting of a few petroglyphs on boulders. At least one historic corral was also located.

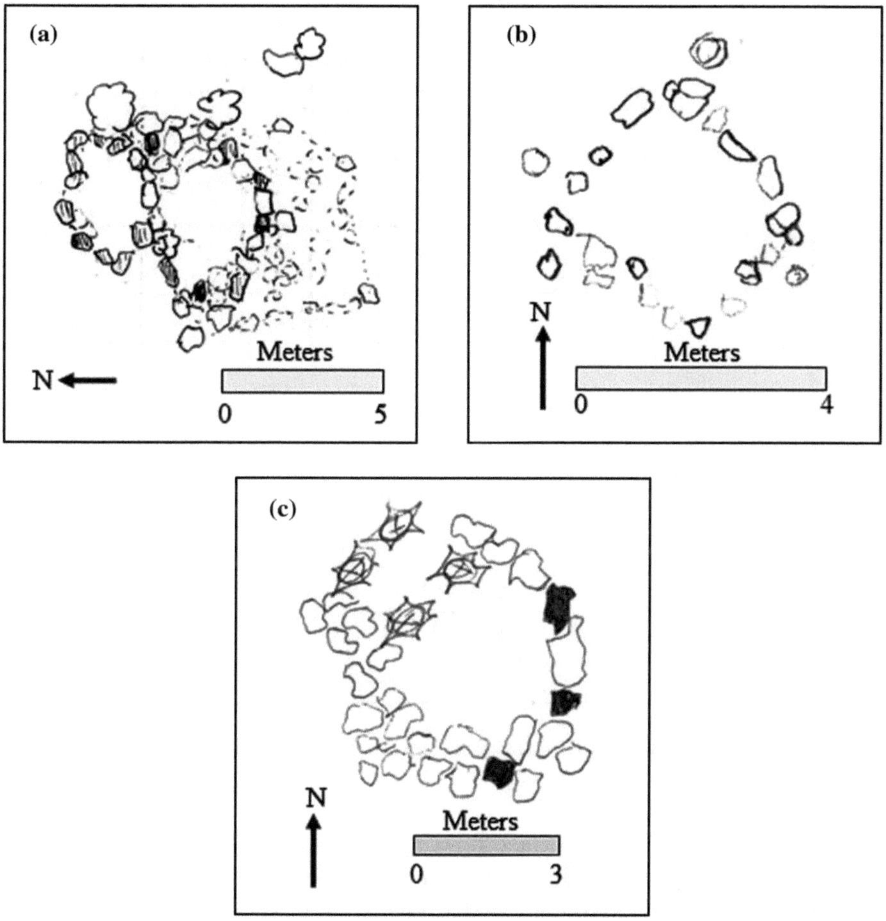

Fig. 3.11 Line drawing showing plan maps of several typical "field house" structures from field notes and coding forms. **a** AZ:A:12:225 (ASM); **b** AZ:A:12:226 (ASM); **c** AZ:A:12:239 (ASM)

3.4.3 Predictive Models of Optimal Maize Field Location

We generated two distinct models, one called the "restrictive" model and the other the "fuzzy logic" model. They are described separately below. For each model we excluded any archeological sites outside the (hydrological) modeling zone; i.e., many sites found in the north half of section 11 for example are not included in the analysis simply because we could not construct complete optimality zones for this area—we did not build the hydrological model outside this small drainage basin.

"Restrictive Model" results. The "restrictive" model shows only those areas where the highest absolute optimality layers intersect. So for example in Fig. 3.13a the light blue pixels are those where only the most optimal zones

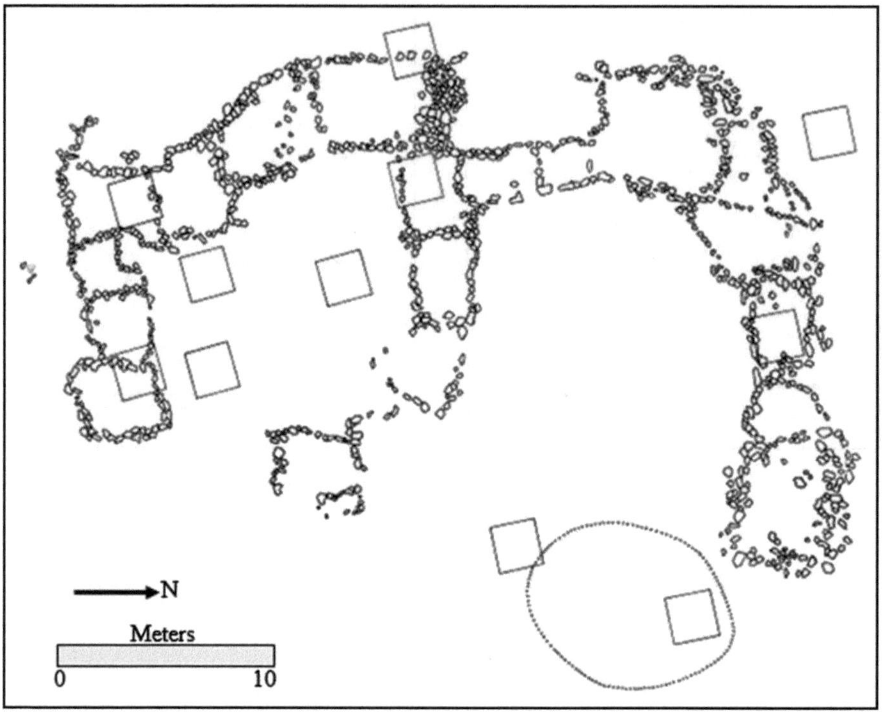

Fig. 3.12 Plan map of "Ken's Big E" one of the larger C-shaped pueblos (AZ:A:12:136 (ASM))

for water infiltration, water surface flow, soils, surface gradient, surface radiant temperature, solar illumination are the highest respective categories. This might be called the "class intersection" model. Only a few pixels (i.e., 30 m × 30 m squares) across the entire study area meet this definition.

The results of the restrictive model are shown in Fig. 3.13. Here the blue area (covering only a few pixels) shows the intersection of only the most optimal conditions from all six layers. The green optimal area contains all the highest ranked layers except for soil—in this case the soil is "moderate" not best. But all other attributes in the "green" model (surface water flow, surface infiltration, surface gradient, solar illumination, and radiant temperature) are highest.

For the yellow areas four attributes were their highest but includes a moderate ranking for soil and water flow depth is poor but infiltration is still highest; orange zone means a soil rating of moderate and poor rating for surface water flow and for infiltration but high rating for all others. These are essentially then four separate discrete models that can be compared across the distribution of site types. The data are shown in Tables 3.2, 3.3, 3.4, 3.5, 3.6.

Let's take the example of the green areas of the restrictive model indicated in Figure 3.13 and Table 3.3. This table shows seven sites of one or two room size and zero scatters within 100 m of a "green" optimality zone (highest optimality

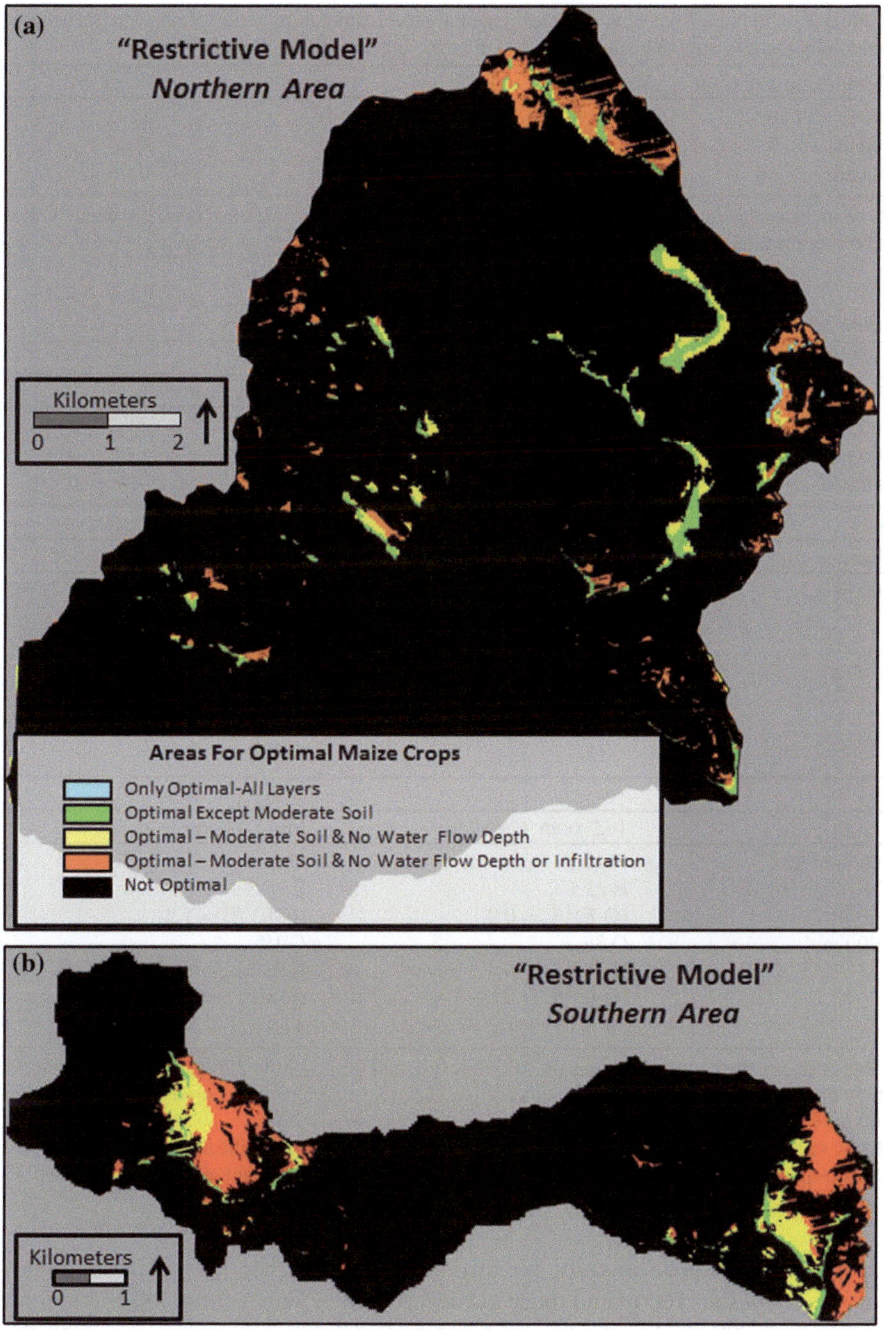

Fig. 3.13 "Restrictive" model results showing northern and southern study areas

Table 3.2 The "blue" restrictive model. There are too many empty cells to calculate statistics—essentially there is only one site within 200 m of the highest optimality area

Distance	1–2 room	Multi-room	Large	Scatter	Rock art
100 m	0	0	0	0	0
200 m	0	1	0	0	0
Outside	36	28	8	15	1

(*Note* includes the entire hydrological model area not just the areas that have been surveyed). Only 0.27 % of the area is within 100 m of the blue optimality zone; 0.93 % of the area is within 200 m

Table 3.3 The "green" restrictive model

A

Distance	1-2 room	Multi-room	Large	Scatter	Rock art
100 m	7	8	0	0	0
200 m	3	9	5	9	0
Outside	27	16	3	7	0

B

Green restrictive	1–2 room + scatter	Multi + large	
100 m or less	O 7 E 8.4 $(O-E)^2/E = 0.2$	O 8 E 6.5 $(O-E)^2/E = 0.3$	15
>100 m	O 46 E 44.5 $(O-E)^2/E = 0.1$	O 33 E 34.5 $(O-E)^2/E = 0.1$	79
	53	41	94

C

	1–2 room + scatter	Multi + large	
200 m or less	O 19 E 23.5 $(O-E)^2/E = 0.9$	O 22 E 17.5 $(O-E)^2/E = 1.2$	41
>200 m	O 36 E 31.5 $(O-E)^2/E = 0.6$	O 19 E 23.5 $(O-E)^2/E = 0.9$	55
	55	41	96

Table A shows the counts, B shows the 100 m result, and C shows the 200 m result. The tables show the whole hydrological model area not just the areas that have been surveyed. Green within 100 m = 13.70 %; within 200 m = 28.77 % of the area
B: Chi square value = 0.7; df = 1, p .05 = 3.84; p. 10 = 2.71 NOT significant at the p = 0.10 level
C: Chi square value = 3.6; df = 1, p .05 = 3.84; p. 10 = 2.71 significant at the p = 0.10 level

except for a moderate in soil). We also combined distance into a simple dichotomy—sites within 100 m and those outside 100 m, or sites within 200 m and those greater than 200 m. The counts are combined in Table 3.3b where it is seen that there are 46 one or two room plus scatter sites father away that 100 m and only seven one-two room plus scatters 100 m or less from a green optimality zone. The Chi square statistic is calculated for a 2 × 2 contingency table with one degree of

Table 3.4 The "yellow" restrictive model

A

Distance	1–2 room	Multi-room	Large	Scatter	Rock art
100 m	8	13	5	4	0
200 m	9	4	2	4	0
Outside	17	12	1	7	1

B

Yellow restrictive	1–2 room + scatter	Multi + large	
100 m or less	O 12 E 17.1 $(O\text{-}E)^2/E = 1.5$	O 18 E 12.9 $(O\text{-}E)^2/E = 2.0$	30
>100 m	O 37 E 31.9 $(O\text{-}E)^2/E = 0.8$	O 19 E 24.1 $(O\text{-}E)^2/e = 1.1$	56
	49	37	86

C

Yellow restrictive	1–2 room + scatter	Multi + large	
200 m or less	O 25 E 27.9 $(O\text{-}E)^2/E = 0.3$	O 24 E 21.1 $(O\text{-}E)^2/E = 0.4$	72
>200 m	O 24 E 21.1 $(O\text{-}E)^2/E = 0.4$	O 13 E 15.9 $(O\text{-}E)^2/E = 0.5$	23
	49	41	95

D

Restrictive for "yellow"	1–2 rooms + scatters	Multi-room + large	
<100 m	O = 12 E = 11.2 $(O\text{-}E)^2/E = 0.06$	O = 18 E = 9.4 $(O\text{-}E)^2/E = 7.87$	30
>100 m	O = 31 E = 31.8 $(O\text{-}E)^2/E = 0.02$	O = 18 E = 26.4 $(O\text{-}E)^2/E = 2.67$	49
	43	36	79

Approximately 16.85 % of the area within the hydrological model is within 100 m of a yellow optimal zone and 29.14 % within 200 m of such a zone. Using only sections 10 and 11 within the hydrological boundary at the "yellow" model. About 26 % of the area is covered by yellow model 100 m radius. Based on proportional area in hydro model boundary with only completely surveyed areas

B: Chi square value = 5.40. Df = 1, p.05 = 3.84; p.10 = 2.71 significant at the p = 0.05 level

C: Chi square value = 1.60; df = 1, p.05 = 3.84; p.10 = 2.71 not significant at the p = 0.10 level

D: Chi square value = 10.62; df = 1; p.05 = 3.84 significant

freedom. Using the distribution of Chi square found in Blalock (1979), it is seen that the observed Chi square value of 0.70 is not statistically significant. To be significant at the 0.10 p value would require that the observed Chi square statistic be 2.71 or larger which it is not. This then indicates there is no pattern in the relationship between site type and distance from green optimality zone.

Table 3.5 The "orange" restrictive model

A

Distance	1–2 room	Multi-room	Large	Scatter	Rock art
100 m	19	22	7	9	0
200 m	6	4	0	5	0
Outside	12	7	1	3	1

B

	1–2 room + scatter	Multi + large	
100 m or less	O 28 E 32.4 (O-E)^2/E = 0.6	O 29 E 24.6 (O-E)^2/E = 0.8	57
>100 m	O 26 E 21.6 (O-E)^2/E = 0.9	O 12 E 16.4 (0-E)^2/E = 1.2	38
	54	41	95

C

	1–2 room + scatter	Multi + large	
200 m or less	O 39 E 40.9 (O-E)^2/E = 0.1	O 33 E 31.1 (O-E)^2/E = 0.1	72
>200 m	O 15 E 13.1 (O-E)^2/E = 0.3	O 8 E 9.9 (O-E)^2/E = 0.4	23
	54	41	95

Note includes the whole hydrological model area not just the areas that have been surveyed. About 26.87 % of the area is within 100 m of an orange optimal area and 42.10 % within 200 m
B: Chi square value = 3.50; Df = 1, p = .05 3.84; p = .10 2.71 significant at the p = 0.10 level
C: Chi square value 0.90; Df = 1, p = .05 3.84; p = .10 2.71 not significant at the p = 0.10 level

The first Tables (such as 3.4b) shows Chi square calculated by constructing the expected frequency by using the marginal totals. For example in Table 3.4b, the observed value is 12 sites (one to two room combined with scatters) found in the hydrological model area within a 100 m radius of modeled yellow area. The expected value is calculated for the cell as (30 × 49)/86 for a value of 17.1. For the cell below it (one or two room sites plus scatters) farther away than 100 m the observed is 37 and the expected is (56 × 49)/86 = 31.9.

Table 3.4d uses a different way to calculate the expected values. Figure 3.14 shows a map of the sections 10 and 11—areas completely surveyed for archaeological sites (the same area as Fig. 3.3c). It is overlain on the hydrological model for the southern study area. The intersection of those two areas results in an area of about 851 acres. Of that area, about 221 acres (or 26 %) is within 100 m radius of a "yellow" optimal field area. Therefore if the distance from a yellow optimal area makes no difference in terms of proximity of field houses to the optimal areas, these site types should be found in the same proportion as the yellow optimal areas in the sub-sampled area. That is, we would expect to find 26 % of each

Table 3.6 Summarizes all Chi square tests of "restrictive" models for distances of 100 and 200 m

Restrictive models	<100 m versus >100 m	<200 m versus >200 m
Blue	n/a	n/a
Green	Chi square value = 0.7 df = 1, p .05 = 3.84; p.10 = 2.71 not significant at the p = 0.10 level	Chi square value = 3.6 df = 1, p .05 = 3.84; p.10 = 2.71 significant at the p = 0.10 level
Yellow	Chi square value = 5.40 Df = 1, p .05 = 3.84; p.10 = 2.71 significant at the p = 0.05 level	Chi square value = 1.60 df = 1, p .05 = 3.84; p.10 = 2.71 not significant at the p = 0.10 level
Orange	Chi square value = 3.50 Df = 1, p = .05 3.84; p = .10 2.71 significant at the p = 0.10 level	Chi square value 0.90 Df = 1, p = .05 3.84; p = .10 2.71 not significant at the p = 0.10 level

There is a significant relationship for the "green" model at a distance of 200 m (significant at the 0.10 level). But this shows there are actually fewer small sites within 200 m than would be expected by chance alone, and more large sites within 200 m. This implies that multi room and larger sites are preferentially closer to green zones than the smaller sites. For the "yellow" the same apparent pattern (significant at the p = 0.5 level). More multi and large sites are found 100 m or less from yellow optimality zones, and fewer 1–2 rooms + scatter sites within 100 m; more small sites outside 100 m and fewer multi plus large sites are farther away. For the orange restrictive model within 100 m there are more multi and large sites and fewer 1–2 room + scatter sites that might be expected if the distribution were random. At a distance of more than 100 m there are more small sites but fewer large sites outside the 100 m zone than would be expected

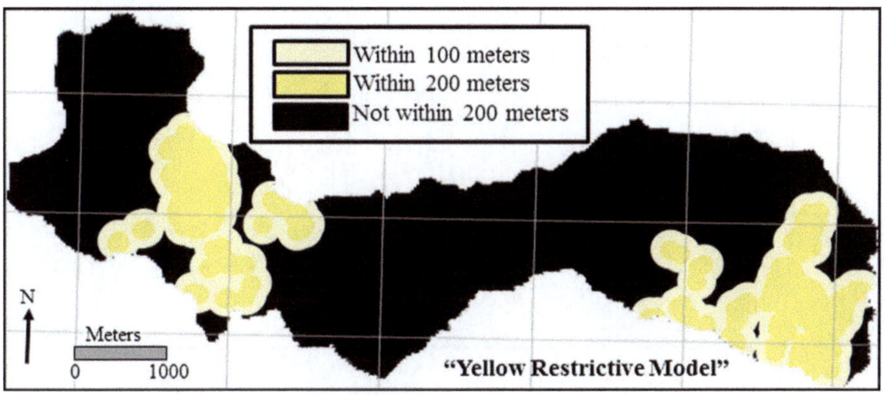

Fig. 3.14 Map of southern study area showing yellow restrictive model optimal areas

site type within 100 m of an optimal field location just by chance if the proportion of sites was the same as the proportion of the area covered by the 100 m optimality zone and buffer for example.

Looking at the Table 3.4d it is clear that as before the distribution of one or two room sites (plus scatters) is essentially unpatterned—there does NOT seem to be any evidence that these types of sites are closer to the yellow optimal areas. On the other hand, there IS a clear pattern of multi and large sites more likely to be near the yellow "optimal" areas. Based strictly on the proportion of the sub- sample within a yellow area (26 %) there are significantly more multi + large sites than might be expected by chance alone.

The "fuzzy logic" model. The fuzzy logic model (Figs. 3.15 and 3.16) on the other hand shows each surface characteristic scaled in a continuous fashion (i.e., from 0–255 for example) and then assigned a color gradation, overlapping with all the other surface characteristics similarly scaled (for detailed explanation of fuzzy logic see Regmi et al. 2010). This model is contingency bound; that is, each unique study area or data set might result in differently defined optimally zones. This is an example of "grouping" rather than classification, since the results are contingency bound; that is, different data sets will result in different groupings (e.g., Dunnell 1971).

Notice that the areas shaded blue in Figs. 3.15 and 3.16 (meaning the most optimal in the fuzzy logic model) are much more common than the blue areas in the restrictive model (Fig. 3.13). The fuzzy logic model emphasizes more strongly the water surface flow layer, implying that water abundance is more important. Since all the attributes are equally weighted, and there are two water attributes, this results in a figure more dominated by blue, all else being equal. Since the solar illumination

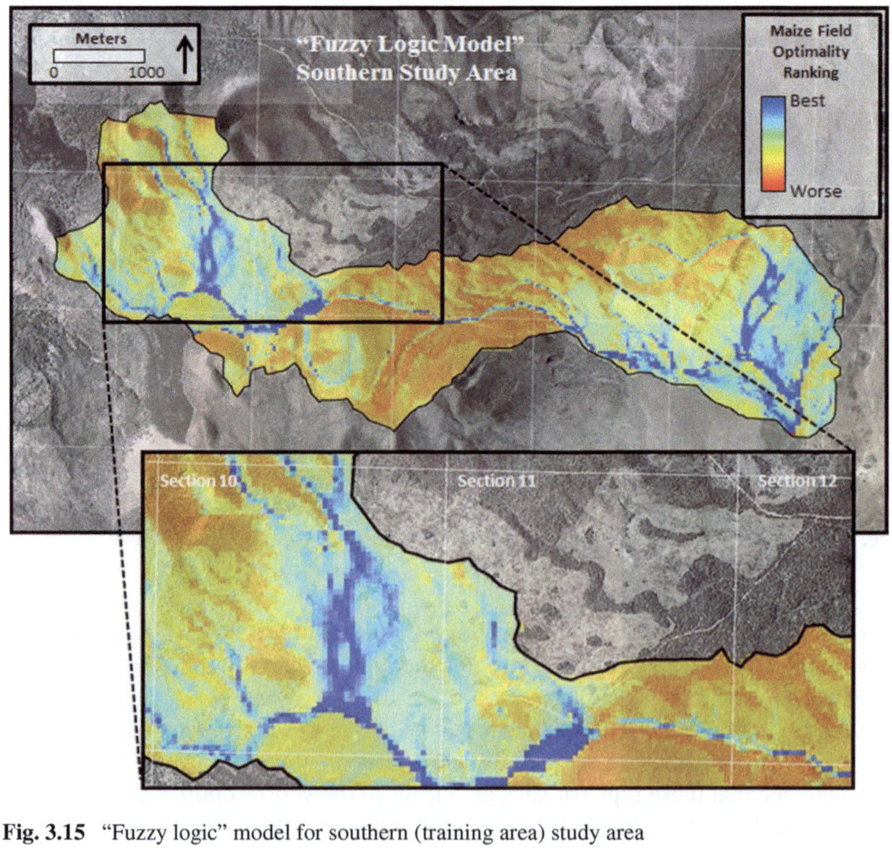

Fig. 3.15 "Fuzzy logic" model for southern (training area) study area

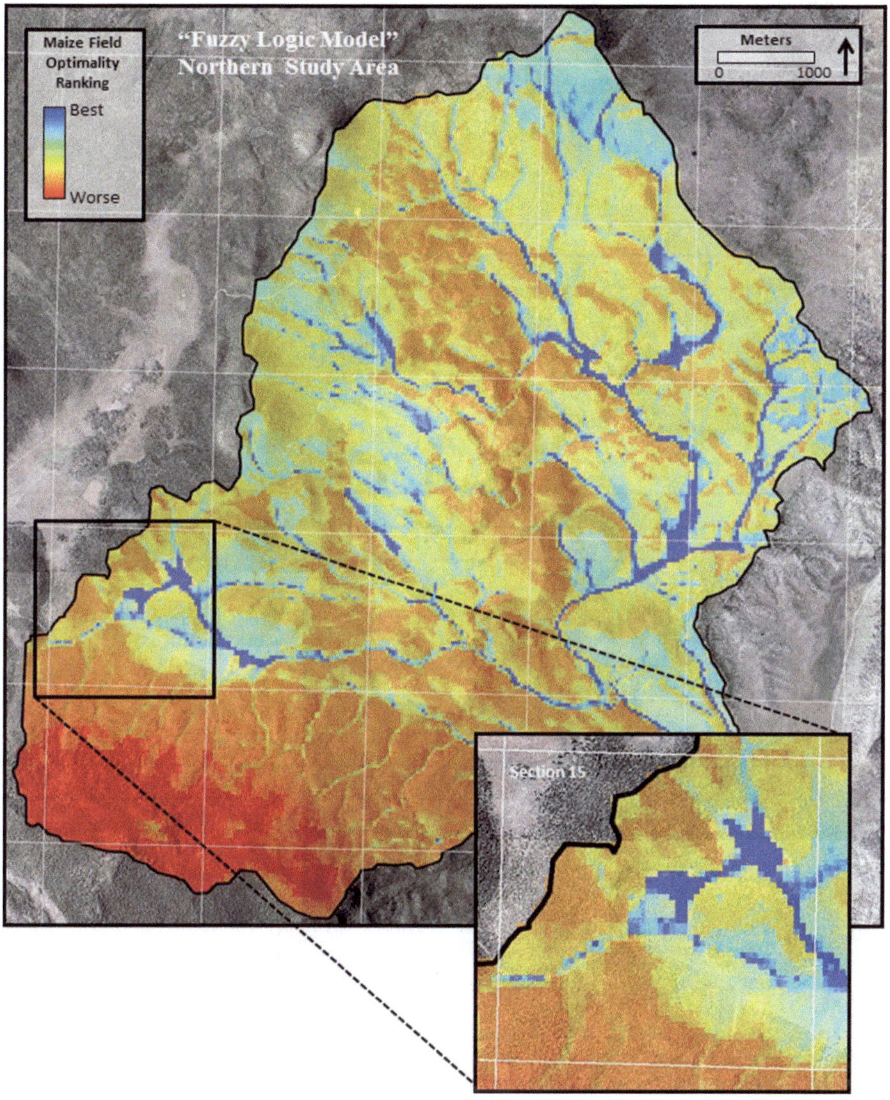

Fig. 3.16 "Fuzzy logic model" model for northern study area

surface gradients and soils are generally (relatively) more homogenous than other attributes, the various shades of blue tend to follow ephemeral drainages.

Tables 3.7, 3.8, 3.9, 3.10 show the results of Chi square testing for the blue, green and yellow "fuzzy logic" optimality modeling. The tables show different site types located within the 100 m radius of a modeled zone (blue, green, yellow); within a 200 m radius; or altogether greater than 200 m radius. The Chi square test statistic was calculated by combining categories of sites and distances to construct a simple

Table 3.7 The "blue" fuzzy logic model

A

	1–2 rooms	Multi-room	Large	Scatters	Rock art	
< 100 m	13	19	7	8	0	47
100-200 m	10	5	1	2	1	19
> 200 m	13	4	1	4	1	23
	36	28	9	14	2	89

B

Fuzzy logic	1–2 rooms + scatters	Multi-room + large	
< 100 m	O = 21	O = 26	47
	E = 27.01	E = 19.99	
	$(O\text{-}E)^2/E = 1.33$	$(O\text{-}E)^2/E = 1.8$	
> 100 m	O = 29	O = 11	40
	E = 22.99	E = 17.01	
	$(O\text{-}E)^2/E = 1.8$	$(O\text{-}E)^2/E = 2.1$	
	50	37	87

C

Fuzzy logic	1–2 rooms + scatters	Multi-room + large	
< 200 m	O = 33	O = 32	65
	E = 37.4	E = 27.6	
	$(O\text{-}E)^2/E = 0.5$	$(O\text{-}E)^2/E = 0.7$	
> 200 m	O = 17	O = 5	22
	E = 12.6	E = 9.4	
	$(O\text{-}E)^2/E = 1.5$	$(O\text{-}E)^2/E = 2.06$	
	50	37	87

D

	1–2 rooms + scatters	Multi-room + large	
< 100 m	O = 20	O = 24	44
	E = 19.4	E = 15.4	
	$(O\text{-}E)^2/E = 0.02$	$(O\text{-}E)^2/E = 4.80$	
> 100 m	O = 24	O = 11	35
	E = 24.6	E = 19.6	
	$(O\text{-}E)^2/E = 0.01$	$(O\text{-}E)^2/E = 3.77$	
	44	35	79

Within the hydrological boundary Approximately 39.76 % of the area is within 100 m of a blue fuzzy logic optimal field location and about 64.9 % is within 200 m, B, C. Fuzzy logic model for blue zone with marginal totals used to calculate Chi square. D. expected values for each cell calculated based on the proportion of areas surveyed within sections 10, 11, and 12 (Fig. 3.2c) within the hydrological model boundary

B: Chi square = 6.03 df = 1, p .05 = 3.84 significant

C: Chi square = 4.76, df = 1, p .05 = 3.84 significant

D: This Chi square test: (1) includes only areas surveyed for archaeology (sec 10, 11, and 12) within the hydrological model area. Total of about 841 acres with ~373 located in the dark blue zone of within 100 m radius of blue optimal area. That's about 44 %. So if the site counts are in same proportion as area about 44 % of the 1–2 room sites should be within dark blue, 44 % of multi room sites in dark blue, etc. Chi square value = 8.60; df = 1; p .05 = 3.84 significant

Table 3.8 The Green fuzzy logic model

A

	1–2 rooms	Multi-room	large	scatters	Rock art	
<100 m	21	22	7	9	1	60
100–200 m	4	6	0	4	1	15
>200 m	12	5	1	3	0	21
	37	33	8	16	2	96

B

	1–2 rooms + scatters	Multi-room + large	
<100 m	O = 30 E = 33.333.3 $(o\text{-}e)^2/e = 0.32$	O = 29 E = 25.7 $(o\text{-}e)^2/e = 0.41$	59
>100 m	O = 23 E = 19.7 $(o\text{-}e)^2/e = 0.55$	O = 12 E = 15.3 $(o\text{-}e)^2/e = 0.7$	35
	53	41	94

C

	1–2 rooms + scatters	Multi-room + large	
<200 m	O = 38 E = 41.16 $(o\text{-}e)^2/e = 0.24$	O = 35 E = 31.84 $(o\text{-}e)^2/e = 0.31$	73
>200 m	O = 15 E = 11.84 $(o\text{-}e)^2/e = 0.84$	O = 6 E = 9.16 $(o\text{-}e)^2/e = 1.09$	21
	53	41	94

B: Chi square = 1.98, df = 1, p .05 = 3.84, p .10 = 2.7 not significant
C: Chi square = 2.48, df = 1, p .05 = 3.84, p.10 = 2.70; not significant

2 × 2 contingency table just like those reported above for the "restrictive" models. The numbers of one or two room sites were combined with scatters; and the multi room sites with large sites. Measures of distance were also dichotomized; either <100 m and greater than 100 m or less than 200 m and greater than 200 m.

Table 3.7d (like 4d above) uses a different way to calculate the expected values. Figure 3.17 shows a map of the "training" area C, although only a portion of this hydrological basin was completely surveyed for archaeological sites (viz., sections 10, 11 and part of 12 as indicated in Fig. 3.3c). The common overlapping area both within the hydrological basin and completely surveyed comes to about 851 acres. Of that area, about 373 acres (or 44 %) is within 100 m radius of a blue optimal field area. Therefore if the distance from a blue optimal area makes no difference in terms of proximity of field houses to the optimal areas, different site types should be found in the same proportion as the blue optimal areas in the sub-sampled area. In this case 44 % of the small sites should be within 100 m of an optimal area, and 44 % of the larger sites, and so on.

Table 3.9 The Yellow fuzzy logic model

A

	1–2 rooms	Multi-room	Large	Scatters	Rock art	
<100 m	27	23	6	13	1	70
100–200 m	9	9	2	5	1	26
>200 m	1	1	0	1	0	3
	37	33	8	19	2	99

B

	1–2 rooms + scatters	Multi-room + large	
<100 m	O = 40 E = 39.84 $(O\text{-}E)^2/E = 0.01$	O = 29 E = 29.16 $(O\text{-}E)^2/E = 0.01$	69
>100 m	O = 16 E = 16.16 $(O\text{-}E)^2/E = 0.01$	O = 12 E = 11.84 $(O\text{-}E)^2/E = 0.01$	28
	56	41	97

C

	1–2 rooms + scatters	Multi-room + large	
<200 m	O = 54 E = 54.23 $(O\text{-}E)^2/E = 0.01$	O = 38 E = 37.77 $(O\text{-}E)^2/E = 0.01$	92
>200 m	O = 2 E = 1.77 $(O\text{-}E)^2/E = 0.03$	O = 1 E = 1.23 $(O\text{-}E)^2/E = 0.04$	3
	56	39	95

B: Chi square = 0.04, df = 1, p .05 = 3.84, not significant
C: Chi square = 0.09, df = 1, p .05 = 3.84; p .10 = 2.70 not significant

Table 3.10 Summary of the outcomes of Chi square for the fuzzy logic models

Fuzzy logic models	<100 m versus >100 m	<200 m versus >200 m
Blue	Chi square = 6.03 df = 1, p .05 = 3.84 significant More multi + large sits close to blue optimal zones than expected	Chi square = 4.76, df = 1, p .05 = 3.84 significant
Green	Chi square = 1.98, df = 1, p .05 = 3.84, p .10 = 2.7 not significant	Chi square = 2.48, df = 1, p .05 = 3.84, p .10 = 2.70; not significant
Yellow	Chi square = 1.98, df = 1, p .05 = 3.84, p .10 = 2.7 not significant	Chi square = 0.09, df = 1, p .05 = 3.84; p .10 = 2.70 not significant

Looking at the Table 3.7d it is clear that as before the distribution of one or two room sites is un-patterned—there does NOT seem to be any evidence that this type of site is closer to "blue" optimal areas. On the other hand, there IS a clear pattern of multi and large sites more likely to be near the blue "optimal" areas. The

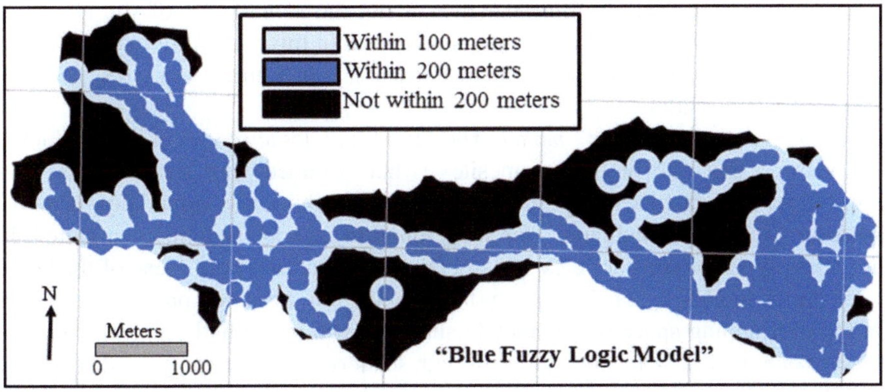

Fig. 3.17 Blue fuzzy logic model for southern area. *Dark blue* indicates areas within 100 m around optimal area; *light blue* encompasses a 200 m radius around such areas

observed number of multi room and large sites is larger than the number expected to be found if they were in the same proportion as the area of the blue optimal zone. Based strictly on the proportion of the sub-sampled area within a blue area (44 %) there are significantly more multi-room and large sites within 100 m of a blue optimality zone than might be expected by chance alone. The expected value for these larger sites is 15.4, but 24 such sites were actually observed.

The "green" fuzzy logic model (Table 3.8) and the "yellow" fuzzy logic model (Table 3.9) indicate the relationship between site types and these optimality zones is essentially random.

3.4.4 Test of Model in New Area; Section 15, T 35 N R8 W

We have examined above the relationship of our modeled optimal maize field locations to the actual distribution of archaeological sites found through intensive pedestrian survey in a small part of the Mt. Trumbull study area (the "training" area, sections 10, 11, and part of 12, T. 34 N R. 8 W). Some of the models (specifically the "blue fuzzy logic" model and the "yellow restrictive" model, both statistically significant) suggest that multi- room and larger sites do tend to be found in close proximity (either less than 200 m or less than 100 m) to some hypothesized optimal field locations. Other results suggest that the smaller sites (the one or two room sites when lumped with scatters) do not tend to be located close to modeled optimal field areas.

We now test whether or not the same pattern holds true for previously unknown areas (the "test" area, viz., section 15 T. 35 N, R. 8 W found to the northwest of Mt. Trumbull). This section was surveyed using the samc archaeological methods as the other portion of the study area. Although the numbers of sites are much lower (i.e., the density of sites per acre is less than a fifth of the site density from the "training" area), we examine whether or not the same kinds of site types are found in similar proportions to the first study area. In this area were recorded three one or two room sites (so called "field houses"), four multi room sites, no large sites, and seven scatters.

Testing "blue fuzzy logic" model. The blue fuzzy logic model in the "training" area showed that one or two room sites (when grouped with scatters) were not found to be associated with optimal areas (their distribution is essentially random). However, multi room and large sites were found to be non-randomly distributed. In fact, these types of sites are clearly associated with optimal field areas of the "blue fuzzy logic" model. In this blue fuzzy logic model, the observed proportion of sites (in the overlapping space of completely surveyed within the hydrological basin) is: 42 % of one or two room sites (grouped with scatters) were found closer than 100 m and 58 % were found further than 100 m; for the multi room grouped with large sites, 70 % were found less than 100 m and only 30 % were greater than 100 m.

Using these proportions as the expected value (E) for a Chi square test of the new area ("test" area, section 15), Table 3.11 shows the calculated Chi square value is 0.29, with df = 1, and the Chi square critical value for p of .05 is 3.84. Although the counts are low, this indicates that the observed value is not significantly different from the expected value, which was calculated as exactly the same proportion found in the "training" area. This indicates that the distribution in the "test" area is essentially the same pattern found previously with the blue fuzzy logic model.

Testing the "yellow restrictive" model. As described above, a statistically significant result was found upon examination of the "yellow restrictive" model for the "training" area (Table 3.4b). In that case we examined site distances to "yellow restrictive" optimal areas at 100 m and greater than 100 m. The calculated Chi square was 5.40. Far more multi and large sites were found at 100 m or less than expected; and fewer one or two room plus scatters were found than expected at this same distance. With df = 1 and a critical p value at .05 of 3.84, this test was significant.

We found that one or two room sites plus scatter were 24 % at 100 m or less and 76 % at greater than 100 m; for multi room plus large it was 49 % 100 m or less and 51 % greater than 100 m. We used these proportions as expected values for Table 3.12 and then examined whether or not the results from the test are significantly different than that found in the "training" area. The calculated Chi square value of 5.52 (with df = 1, and p. critical at .05 of 3.84) indicates that there is statistically significant difference between the "training" and test area for the yellow restrictive model.

Table 3.13 shows a different way to calculate whether or not sites in the test area are randomly distributed or not. Like earlier tables (viz, Tables 3.4b, 3.7b for example) we used the more conventional marginal totals to calculate the expected values for the contingency table. The calculated Chi square of 1.78 (df = 1, critical Chi square value at p = .05 of 3.84) shows that the pattern is not significantly different than random. This test again suggests that the pattern seen in the "training" area for the yellow restrictive model is not repeated in the "test" area.

Table 3.11 Blue fuzzy logic for northern study area (the test area, Fig. 3.2b)

	1–2 room + scatters	Multi room + large	
<100 m	O = 5 E = 4.2	O = 4 E = 4.2	9
>100 m	O = 5 E = 5.8	O = 2 E = 1.8	7
	10	6	16

Chi square calculated = 0.29, df = 1, Chi square critical at p 0.05 = 3.84; not significant

Table 3.12 The "yellow restrictive" model for northern study area results

	1–2 room + scatters	Multi room + large	
<100 m	O = 0 E = 2.4 (O−E)2/E = 2.4	O = 1 E = 2.94 (o−e)2/E = 1.28	1
>100 m	O = 10 E = 7.6 $(O-E)^2/E = 0.64$	O = 5 E = 3.06 $(O-E)^2/E = 1.20$	15
	10	6	16

This test uses as the "expected values" the actual percentages taken from the yellow restrictive model in the southern area. i.e., 24 % of sites less than 100 m from those optimal areas were 1–2 rooms or scatters; and 76 % were found farther away than 100 m
Calculated Chi square = 5.52, df = 1, critical Chi square at p 0.05 = 3.84 significant

Table 3.13 The "yellow restrictive" model for northern study area results, this time testing against random

	1–2 room + scatters	Multi room + large	
<100 m	O = 0 E = 0.625 (O−E)2/E = 0.625	O = 1 E = 0.375 (o−e)2/E = 1.04	1
>100 m	O = 10 E = 9.375 $(O-E)^2/E = 0.04$	O = 5 E = 5.625 $(O-E)^2/E = 0.07$	15
	10	6	16

This test uses as the "expected values" calculated the conventional way, i.e., row total times column total divided by grand total to get the expected values if sites were randomly distributed
Calculated Chi square = 1.78, df = 1, critical Chi square at p 0.05 = 3.84 not significant)

3.5 Discussion and Conclusions

The Mt. Trumbull area of the Uinkaret Plateau is a part of the Colorado Plateau long inhabited by agriculturalists. In this area maize agriculture was a very important element of subsistence and was based on rainfall agriculture, not irrigation. Recent investigations have shown dozens if not hundreds of new archaeological sites—some relatively large C and L-shaped pueblos of ~20 rooms—and dozens of smaller sites which in many places in the Southwestern US would comfortably be labeled "field houses."

Unlike other areas of the Southwest where such structures are associated with clear agricultural features such as check dams, rock mulch gardens, or gridded fields, there are virtually no clues remaining in Mt. Trumbull about the functions and purposes of these "field houses." It is presumed by some that these structures mark the locations of "fields" but this argument is largely tautological. In this project we attempt to identify abiotic factors that when combined, can help us identify where the "best" or optimal places may be on the landscape for productive agricultural fields. We use variables such as modeled rainfall intensity and infiltration, soil type, slope, air temperature, radiant surface temperature (from satellite data), illumination, and soil moisture. Simply put, those places with the best soils, most soil moisture at critically important times of the year for maize, and the longest growing season are suggested to be "optimal" for maize productivity.

We constructed two types of optimality models which we call above the "restrictive" model and the "fuzzy logic" model. We examined initially whether different types of sites are found "close" (either 100 m or 200 m) to an optimal place for maize agriculture, or father away (generally greater than 200 m but could also mean greater than 100 m). The two predictive models (restrictive and fuzzy logic) generally agree that smaller sites (one or two room structures and scatters) are NOT located preferentially close to our optimal areas. Conversely, and somewhat unexpectedly, there IS a clear preference for larger sites to be found closer to optimal areas.

In particular, in the "training" area for the "green" restrictive model there are a disproportionate number of large sites found within 200 m of the green optimal zone; for the yellow optimal zone there is a statistically significant relationship between larger sites and the yellow zones at 100 m or less. For the blue fuzzy logic model, there is a strong relationship between the number of large sites and a blue zone both at the 100 m and 200 m distance. But it seems as though smaller sites ("field houses") are randomly distributed across the landscape—at least they are not found closer to our modeled optimal areas.

The smaller sites (one or two room structures) do not seem to be "field houses" if by field houses we mean the kinds of sites near fields (as we have defined them), thought by some at least to have been primarily used to monitor agricultural fields and protect crops from invaders, pests, etc. Using the proportion of site types from the "training" area, we performed several Chi square tests using those proportions against the actual values found in a previously unknown area (test area Fig. 3.3b). It is true that the quantities of sites are much lower in this area than in the test area, making the numbers much lower. It was found that the proportions of large sites close to the fuzzy logic blue optimally zone in the "test" is indistinguishable from the "training" area, meaning essentially the same pattern is found in section 15; viz., there are disproportionally more large sites found closer to blue optimal areas in the fuzzy logic model than would be expected by chance alone.

We tested whether or not these small sites were located near the most optimal areas for maize agriculture. We did not test how close these sites were to the less optimal zones; we are able to say however that these small sites are not located close to the most optimal places as might be expected if they are in fact "field houses" (sensu Haury 1956 and others). It may well be that these smaller sites were established after Pueblo II (PII) population growth had created the larger C- and L-shaped pueblos (through accretion?), and that these smaller 1–2 room structural sites did in fact act as field houses—but in more marginal locations. As this portion of the Mt. Trumbull area became increasingly "packed" during the PII period, it may be that that kin groups (from any nearby valley—not necessarily the closest large habitation) established field houses to monitor their more marginal fields. This process might have intensified in the 13th century as environmental conditions deteriorated, or at any time when summer monsoonal flows became reduced for long periods.

These first efforts in applying a multi-layered modeling approach for the prediction of ancient crop locations are encouraging. Future efforts will include testing the models in larger areas of Mt. Trumbull where the site counts are larger. Additionally, we intend to expand to areas outside of the immediate Mt. Trumbull; area perhaps elsewhere on the Uinkaret Plateau. Also, we may be able to further constrain the relationship between ASTER radiant temperature at different times of day and air temperature, by examining higher-temporal resolution ASTER thermal data. This is especially important for accurate modeling of the length of the growing season. Future work should also incorporate a model that considers the effect of cold air sinks upon growing season duration. Finally, a critical step is to develop a regional chronology to better partition the archaeological record as we need to better understand the occupational history of the region to determine if these smaller structural sites are contemporaneous with larger habitations.

References

Altschul JH, Fairley HC (1989) Man, models, and management: An overview of the Arizona strip and the management of its cultural resources. Prepared for the USDA Forest Service and USDI Bureau of Land management by Statistical research, Plateau Archaeology and dames & Moore, Inc

Anschuetz K, Maxwell T (1986) A multidisciplinary investigation of prehistoric Puebloan gardens in the lower Rio Chama valley, New Mexico. In: Papers presented at the 9th annual ethnnobiology conference, Society of ethnobiology, University of New Mexico, Albuquerque, NM, 21–23 March 1986

Arya LM, Paris JF (1981) A physiochemical model to predict the soil moisture characteristics from particle-size distribution and bulk density data. Soil Sci Soc Am J 45:1023–1030

Arya LM, Leij FJ, van Genuchten MT, Shouse PJ (1999) Scaling parameter to predict the soil water characteristic from particle-size distribution data. Soil Sci Soc Am J 63:510–519

Billingsley GH, Hamblin K (2001) Geological map of part of the Uinkaret Volcanic field, Mohave County, Northwestern Arizona. USGS MF-2368

Blalock HM (1979) Social statistics, Revised 2nd Edn. McGraw Hill Publishing, New York

Bradfield M (1971) The changing pattern of Hopi agriculture. Royal anthropological Institute of Great Britain and Ireland. In: Royal anthropological institute occasional paper 30, London

Bryan K (1929) Floodwater farming. Geogr Rev 19:441–456

Buck PE (2002) Preliminary Report of Archaeological Research at Uinkaret Pueblo, AZ:A:12:14 (MNA), Prepared by Paul E. Buck, Ph.D. (with ceramic analysis by Laureen Perry) Desert Research Institute

Buck PE (2005) Results of field investigations in the Grand Canyon-Parashant National Monument during the summer 2004. Report prepared for John Herron under Permit AZ000279 May 2005

Buck PE (2006) Report of the 2005 Season, Mt. Trumbull prehistory project. Report prepared for John Herron, GC-PNM, under BLM permit AZ0002789. April, 2006

Buck PE, Sakai S (2005) The Mt Trumbull prehistory project, Society for American archaeology annual meeting, Salt Lake City, 2005, program with abstracts

Buck PE, Sakai S (2008) The Ancestral Puebloan Occupation of the Mt Trumbull area, Uinkaret Plateau, Grand Canyon-Parashant National monument. In: Abstracts of the 73rd annual meeting of the Society for American archaeology, Vancouver, Canada

Buck PE, Martin C, Sakai S (2004) Surface collection and limited soil probing at the zip code site (AZ:A:12:131 [BLM]). Report prepared for John Herron, GC-P NM under permit AZ000248

Clark County Regional Flood Control District (CCRFCD) (1999) Hydrologic criteria and drainage design manual. Las Vegas, NV

Cook ER, Esper J, RD D'Arrigo (2004) Extra-tropical Northern Hemisphere land temperature variability over the past 1000 years. Quat Sci Rev 2:2063–2074

Cordell L (2009) Archaeology of the Southwest, 2nd Edn. Left Coast Press, Walnut Creek

Dominguez S, Kolm KE (2005) Beyond water harvesting: a soil hydrology perspective on traditional southwestern agricultural technology. Am Antiq 70(4):732–765

Dorshow WB (2012) Modeling agricultural potential in Chaco Canyon during the Bonito phase: a predictive geospatial approach. J Archaeol Sci 39:2098–2115

Dunnell RC (1971) Systematics in prehistory. Macmillan (The Free Press), New York

Ellis F (1978) Small structures used by historic Pueblo peoples and their immediate ancestors. In: Ward A (ed) Limited activity and occupation sites: a collection of conference papers. Center for Anthropological Studies, Albuquerque, pp 59–68

Hack JT (1942) The changing physical environment of the hopi Indians of Arizona. Papers of the Peabody Museum of American archaeology and Ethnology, Harvard. Vol. Xxxv no. 1

Haury E (1956) Speculations on prehistoric settlement patterns in the Southwest. In: Willey GR (ed.) Prehistoric settlement patterns in the new world. Viking Fund Publ. Anthropology, vol 23, pp 3–10

Hendricks DM (1985) Arizona soils. University of Arizona, Tucson

Herhahn C (1995) 14th century dry farming features in the Northern Rio Grande Valley, New Mexico. In: Toll HW (ed) Soils, water, biology and belief in prehistoric and traditional southwestern agriculture. New Mexico Archaeological Council Special Pub 2 (77–84)

Honeycutt L (1995) Dryland gardening in southwest Colorado: past and present. In: Toll HW (ed) Soil, water, biology and belief in prehistoric and traditional southwestern agriculture. New Mexico Archaeological Council Special Pub 2 (369–373)

Huete AR, Jackson RD (1987) Suitability of spectral indices for evaluating vegetation characteristics on arid rangelands. Rem Sens Envir 23(2):213–232

Jorgensen W (2004) Soil survey of Mohave County area, Arizona, northeastern part, and part of Coconino County/United States Department of Agriculture, Natural Resources Conservation Service… [et al.]; in cooperation with the Arizona Agricultural Experiment Station and the Kaibab-Paiute Tribe

Kohler TA (1992) Field houses, villages, and the tragedy of the commons in the early northern Anasazi Southwest. Am Antiq 57:617–635

Lang R (1995) The fields of san Marcos: agriculture at a great town of the Galisteo basin, Northern new Mexico. In: Toll HW (ed) Soil, water, biology and belief in prehistoric and traditional southwestern agriculture. New Mexico Archaeological Council Special Pub 2 (41–76)

Leonard RD (1989) Resource specialization, population growth, and agricultural production in the American Southwest. Am Antiq 54(3):491–503

Lyneis M (1995) The Virgin Anasazi, Far Western Puebloans. J World Prehist 9:199–241

Martin CM (2009) Analysis of flaked stone Lithics from Virgin Anasazi sites near Mt. Trumbull, Arizona Strip. Unpublished Master's thesis, Department of Anthropology and Ethnic Studies, University of Nevada, Las Vegas

Mein RG, Larson CL (1973) Modeling infiltration during a steady rain. Water Resour Res 9(2):384–394

Meyer WJ, Ahmad S, Young MH, Shafer DS, Miller JJ Chief K (2010) Effect of spatial and temporal variability of antecedent moisture content on model-generated runoff from an arid watershed. Masters Thesis, University of Nevada, Las Vegas

Mindeleff V (1891) A study of Pueblo architecture in Tusayan and Cibola. Washington. Government Printing Office, 1891. (**REPRINT*)

Mualem Y (1976) A new model for predicting the hydraulic conductivity of unsaturated porous media. Wat Resour Res 12(3):513–522

Moffitt K, Chang C (1978) The mount trumbull survey. W Anasazi Rep 1:185–250

Moore BM (1978) Are Pueblo field houses a function of urbanization? In: Ward A (ed) Limited activity and occupation sites; a collection of conference papers. Center for Anthropological Studies, Albuquerque, pp 9–16

Neuman SP (1976) Wetting front pressure head in the infiltration model of Green and Ampt. Water Resour Res 12(3):564–566

O'Brien JS, Jorgensen GR, Garcia R (2007) FLO-2D Software Version 2007.6. Nutrioso, AZ

Parasuraman K, Elshorabagy A, Si BC (2006) Estimating saturated hydraulic conductivity in spatial variable fields using neural network ensembles. Soil Sci Soc Am J 70:1851–1859

Preucel RW (1988) Seasonal agricultural circulation and residential mobility; a prehistoric example from the Pajarito plateau, New Mexico. Unpublished PhD dissertation, UCLA

Rawls WJ, Brakensiek DL, Savabi MR (1989) Infiltration parameters for rangeland soils. J Range Manag 42:139–142

Regmi NR, Giardino JR, Vitek JD (2010) Assessing susceptibility to landslides: using models to understand observed changes in slopes. Geomorphology 122:25–38

Reid JJ, Whittlesey S (1997) The Archaeology of Ancient Arizona. University of Arizona Press, Tucson

Rhode D (1995) Estimating Agricultural carrying capacity in the Zuni region, west-central New Mexico: a water allocation model. In: Toll HW (ed) Soils, water, biology and belief in prehistoric and traditional southwestern agriculture. New Mexico Archaeological Council Special Pub 2 (85–100)

Russell S (1978) The agricultural field house: a Navajo limited occupation and special use site. In: Ward A (ed) Limited activity and occupation sites; a collection of conference papers. Center for Anthropological Studies, Albuquerque

Sakai S (2005) Change in the clay procurement in the ceramic production among the highland virgin branch. In: Abstracts of the 70th annual meeting of the society for American Archaeology Salt Lake City, p 256, UT March 30–April 3, 2005

Sakai S (2007) Explaining change in production and distribution pattern of olivine ceramics in the Arizona strip of Northern Arizona. In: Abstracts of the 72th annual meeting of the society for American Archaeology, Austin TX, p 363

Sakai S (2010) Change in production and distribution pattern of olivine tempered ceramics in the Arizona strip and adjacent areas in the American Southwest. In: Abstracts of the 75th annual meeting of the society for American Archaeology St. Louis Missouri, p. 215, April 14–18, 2010

Sakai S (2012) Change in clay sources of olivine-tempered ceramics in the Arizona strip and adjacent areas in the American Southwest. In: Abstracts of the 77th annual meeting of the society for American Archaeology Memphis Tenn, p. 306, 18–22 April 2012

Schaap MG, Leij FJ, van Genuchten MTh (2001) Rosetta: a computer program for estimating soil hydraulic parameters with hierarchical pedotransfer functions. J Hydrol 215:163–176

Toll HW (1995) Soil, water, biology and belief in prehistoric and traditional southwestern agriculture. In: Toll HW (ed) New Mexico Archaeological Council Special Pub 2 (41–76)

Upham S (1984) Adaptive diversity and Southwestern abandonment. J Anthropol Res 40(2):235–256

van Genuchten MT (1980) Closed-form equation for predicting the hydraulic conductivity of unsaturated soils. Soil Sci Soc Am J 44(5):892–898

Woodhouse CA, Meko DM, MaCDonald GM, Stahle DW, Cook ER (2010) A 1,200 year perspective of 21st century drought in southwestern North America. Proc Natl Acad Sci 107:21283–21288

MIX
Papier aus verantwortungsvollen Quellen
Paper from responsible sources
FSC® C105338

If you have any concerns about our products,
you can contact us on
ProductSafety@springernature.com

In case Publisher is established outside the EU,
the EU authorized representative is:
Springer Nature Customer Service Center GmbH
Europaplatz 3, 69115 Heidelberg, Germany

Printed by Libri Plureos GmbH
in Hamburg, Germany